BLUE RIDGE FIRE TOWERS

BLUE RIDGE FIRE TOWERS

ROBERT SORRELL

THE History PRESS

Published by The History Press
Charleston, SC 29403
www.historypress.net

First published 2015

Manufactured in the United States

ISBN 978.1.62619.817.3

Library of Congress Control Number: 2014956811

Notice: The information in this book is true and complete to the best of our knowledge. It is offered without guarantee on the part of the author or The History Press. The author and The History Press disclaim all liability in connection with the use of this book.

CONTENTS

ACKNOWLEDGEMENTS

I want to thank the countless organizations, government agencies and people who have graciously contributed to this book, including, in no particular order, Lu Ellsworth of the High Knob Enhancement Corporation; Clinch Ranger District staff of the George Washington and Jefferson National Forest; Rabun County, Georgia Historical Society; Towns County, Georgia Historical Society; Union County, Georgia Historical Society; the Town of Unicoi, Tennessee; staff at the Great Smoky Mountains National Park; staff at the University of Tennessee Library and Great Smoky Mountains Regional Project; Charlie Roth, grandson of photographer Albert "Dutch" Roth; Friends of Great Smoky Mountains National Park; Tennessee Supreme Court justice Gary Wade; architect Tom Trotter; the staff at the *Erwin Record* newspaper; Phyllis Cline of Elizabethton, Tennessee; Betty Hodge of Bristol, Tennessee; Lisa Kestner Quigley of Hayters Gap, Virginia; Ron Stafford, Keith Argow and Krissy Reynolds, members of the Forest Fire Lookout Association; Tennessee State Library and Archives; National Register of Historic Places; Beth Bradford of New Jersey; the Morgan family of Bakersville, North Carolina; Doris Smithson Temple; Gale Colbaugh of Elizabethton, Tennessee; Rondall Ellis of Elizabethton, Tennessee; Lloyd Allen of Burnsville, North Carolina; Ted Wisniewski; Highlands, North Carolina Historical Society; Annie Laurie McDonald at the North Carolina Historic Preservation Office; Keith Peters; Interpretive Ranger Cathy Taylor at Paris Mountain State Park in South Carolina; staff at the South

Carolina Forestry Commission; Virginia Department of Forestry; Friends of Cherokee National Forest; and Tusquitee Ranger District staff of the Nantahala National Forest.

I also want to thank the numerous people who have called me and sent e-mails and other correspondence with information and histories of fire towers. Lastly, thanks to the staff at The History Press for their support of this project.

INTRODUCTION

Keeping a watchful eye over the Blue Ridge Mountains in the eastern United States for more than a century, fire lookout towers have become icons in America's great wild landscape. After decades of devastating wildfires that consumed millions of acres of forestland and communities across the United States during the late nineteenth and early twentieth centuries, conservationists created a network of fire lookout towers to protect the woodlands. They were designed so trained men and women, who manned the towers for days and weeks at a time, could spot flames and smoke from miles away. Using special fire finding equipment, such as the Osborne fire finder, lookouts would work together to pinpoint a fire's location and communicate with headquarters to dispatch firefighters.

Working in solitude and sometimes high above the land in a small seven- by seven-foot wood box-like structure, lookouts became the eyes of the state and federal forestry departments. Lookout employees were, and still are, brave, smart and self-sufficient. They were also rapid and precise in spotting smoke, so firefighters could quickly venture into the mountains and douse the flames. They worked alone, not only watching for fires but also recording weather data, maintaining communications and greeting random hikers and tourists.

Lookout platforms, from which employees watch for smoke, come in all shapes and sizes. The earliest lookouts, those built before and shortly after 1900, included mountaintop tents, cabins and makeshift, temporary towers. Wood, steel and stone towers, typical of lookouts remaining today in the Blue Ridge, came later, primarily during the Great Depression era of the 1930s. Sometimes lookouts required no structure at all. A man or woman would simply stand at the highest point on a mountain and watch for smoke or flames with a pair of binoculars and the means to communicate with firefighters. Some lookout structures were also built in trees. A ladder would be attached to the side of a tall tree, and a seat or platform would be constructed near the tree's top.

Over the years, temporary towers have been replaced with permanent towers. A small makeshift tower would be torn down and replaced with a large steel tower. In many cases, especially since the late twentieth century, towers have been taken down and replaced with nothing at all.

And while lookouts come in all shapes and sizes, many of the structures in the Blue Ridge are tall, sleek towers, rising above the balds and ridges of the mountains. These are the structures that give visitors with weak stomachs and frail nerves a case of the heebie-jeebies. Connie Miller of Blount County, Tennessee, recalled a visit to the old steel Look Rock Tower in the Great Smoky Mountains back in the late 1950s. She made the mistake of climbing the tower and looking back down from the lookout's cab. She froze. Thankfully, she was not alone, and her friends helped her back down the tower.

Government agencies, conservation organizations, businesses and private individuals have built thousands of lookout towers across the country. Idaho, with nearly one thousand towers, had more than any other state. According to the U.S. Forest Service, the state had fewer than two hundred towers in 2014. Kansas is the only state to have never had a lookout tower. Most towers built in the United States were constructed between the 1930s and 1940s. There were lookouts in the late nineteenth century, but none remains today. Lookout towers, although not generally used for fire watching, are still built today, if only for tourism.

The first known tower for forest fire detection in the United States was built in 1902 on private lands in Idaho. The Forest Lookout Museum said that the 1902 site was a woman who climbed a makeshift ladder and sat on a big tree limb at Bertha Hill, Idaho.

An even earlier fire lookout was constructed in Helena, Montana, but it was especially built to warn citizens of fires in the town. The "Guardian of the Gulch" Fire Tower was originally built in 1874 and had a bell to warn citizens of fire. The tower still stands over Helena and has had several repairs over the years.

The U.S. Forest Service built its first fire tower in 1915 in Oregon. The first tower in the eastern United States was built in the Adirondack Mountains of New York in 1909. The first full-time female fire lookout was hired in 1914.

Although fire lookouts were used as early as 1900 in the Blue Ridge Mountains, towers started to be constructed in the 1910s. The Civilian Conservation Corps (CCC) and the Works Progress Administration (WPA), two post–Great Depression government agencies, built a vast majority of the towers in the Blue Ridge and elsewhere in the country. Young unemployed men enlisted by the thousands, lived in CCC camps, similar to military barracks, and worked hard labor. The CCC worked with the U.S. Forest Service, the National Park Service, the Tennessee Valley Authority (TVA) and the Bureau of Land Management on various projects, including the construction of fire towers and firefighting.

By the late twentieth century, however, new technology, such as airplanes, surveillance cameras and cellular phones had begun to change the way foresters watched for fires, and the nation began to decommission its fire towers. One by one, the government took towers out of service. Only a few towers remain manned, and most of those are in the western United States. Fewer than a dozen towers are still manned in the eastern United States. Towers that are no longer used to detect fires have been abandoned and dismantled. Remaining towers have deteriorated and have become liabilities for states and the federal government.

Many abandoned and obsolete towers continue to be removed from the forests, yet there are organizations, such as the Forest Fire Lookout Association (FFLA), and individuals, like Lloyd Allen of Burnsville, North Carolina, that fight for the retention of towers. Allen helped save the Little Snowball Fire Tower in the 1980s and eventually rebuilt it in the Barnardsville community. He bought the tower for $300 in 1980. It was relocated to the Big Ivy Historical Park in 2005. He is now seeking protection of the remaining towers in the Pisgah and Nantahala National Forests, especially Green Knob, where his father-in-law worked.

A handful of the still-standing towers have been restored for visitors, but many have not. Lookout towers at Pinnacle Mountain, Mount Cammerer and Panther Top have been saved and preserved. Some towers have also been rebuilt, like the High Knob Fire Tower in Wise County, Virginia, which was destroyed by arsonists in 2007. A new fireproof High Knob Fire Tower was reopened to the public in 2014.

And then there are towers, like Mendota and Hayters Knob in southwest Virginia, that continue to deteriorate. The two steel lookouts are located at different sites on Clinch Mountain in Washington County. Both towers have previously been slated for removal but have been saved by local residents, fire lookouts and historians. Hayters Knob is located in the Channels State Forest, which will likely protect the tower, despite its deteriorating structure.

CHAPTER 1
EARLY LOOKOUTS

In the early days of forest fires, they were considered simply and solely as acts of God, against which any opposition was hopeless and any attempt to control them not merely hopeless but childish. It was assumed that they came in the natural order of things, as inevitably as the seasons or the rising and setting of the sun. Today, we understand that forest fires are wholly within the control of man.
—*Gifford Pinchot,* The Fight for Conservation, *1910*

Early American conservationist Gifford Pinchot, the country's first forestry chief, wrote about wildfires in his 1910 book, *The Fight for Conservation.*

He fought vigorously for the protection of the nation's forests, and he believed the government was responsible for managing the wildlands and controlling fires. And if an unwanted fire ignited in the forest, Pinchot supported suppressing the flames. Before Pinchot became the face of conservation in America, people thought that man had no control over fires. Citizens and government officials alike believed that fires were a force of nature that could not be abated or prevented.

Pinchot, a lifelong forest enthusiast, and other conservationists worked with private organizations and the federal government to manage the nation's forests, primarily by purchasing lands and fighting uncontrollable fires.

In 1909, Pinchot wrote that "the best way for the government to promote each of these three great uses [which included irrigation, timber, and grazing] is to protect the forest reserves from fire." Although Pinchot did not necessarily believe that all fires should be stamped out, he favored suppression of wildfires, especially those that did not support healthy forest growth. Pinchot's management beliefs were similar to those used today by the National Forest Service in that controlled and prescribed fires are considered healthy.

Pinchot supported efforts to develop a network of lookout towers to protect the forests. By the mid-twentieth century, thousands of fire towers had been constructed across the country, from Maine to Hawaii. In the Blue Ridge Mountains, where Pinchot first introduced modern forest management, the Depression-era Civilian Conservation Corps built hundreds of wood and steel towers.

Pinchot, born in 1865 in Connecticut, had always been fond of the forests. His family owned a large timber company, but he preferred conservation of the woods. He studied forestry in Nancy, France, before returning to work in the United States. In the early 1890s, George Washington Vanderbilt hired Pinchot to manage the forests at Biltmore Estate in western North Carolina. Pinchot's introduction of a new modern forest management system at Biltmore is considered the birth of forest conservation in America. Pinchot managed about 125,000 acres of forestland south of Asheville.

In 1898, Pinchot, who made a name for himself at Biltmore, became head of the federal government's forestry department under President Theodore Roosevelt. At the time, the new forestry department was in its infancy. Pinchot helped Roosevelt, whose presidential legacy is built on conservation, develop the agency. By 1905, management of the forestry department had transferred from the Department of Interior to the Department of Agriculture. Pinchot was named chief of the newly established National Forest Service, which is the name of the nation's present forestry agency.

In early 1910, President William Howard Taft fired Pinchot—widely believed to be due to a politically motivated land dispute in Alaska. Pinchot went on to become governor of Pennsylvania.

Throughout his forestry and political career, and in retirement, Pinchot fought for the protection of the nation's woodlands. Pinchot's successors—especially Henry Graves, William B. Greeley, Robert Stuart, Ferdinand Silcox and Earle Clapp—continued to fight for the nation's forests. Each chief, however, brought his own beliefs in forest management and the importance of fire.

Pinchot's first successor, Henry Graves, said in 1913 of fire detection and suppression, "It is the fundamental obligation of the Forest Service and takes precedence over all other duties and activities." Graves and Silcox believed in complete fire suppression, which Pinchot became a vocal critic of during the early twentieth century.

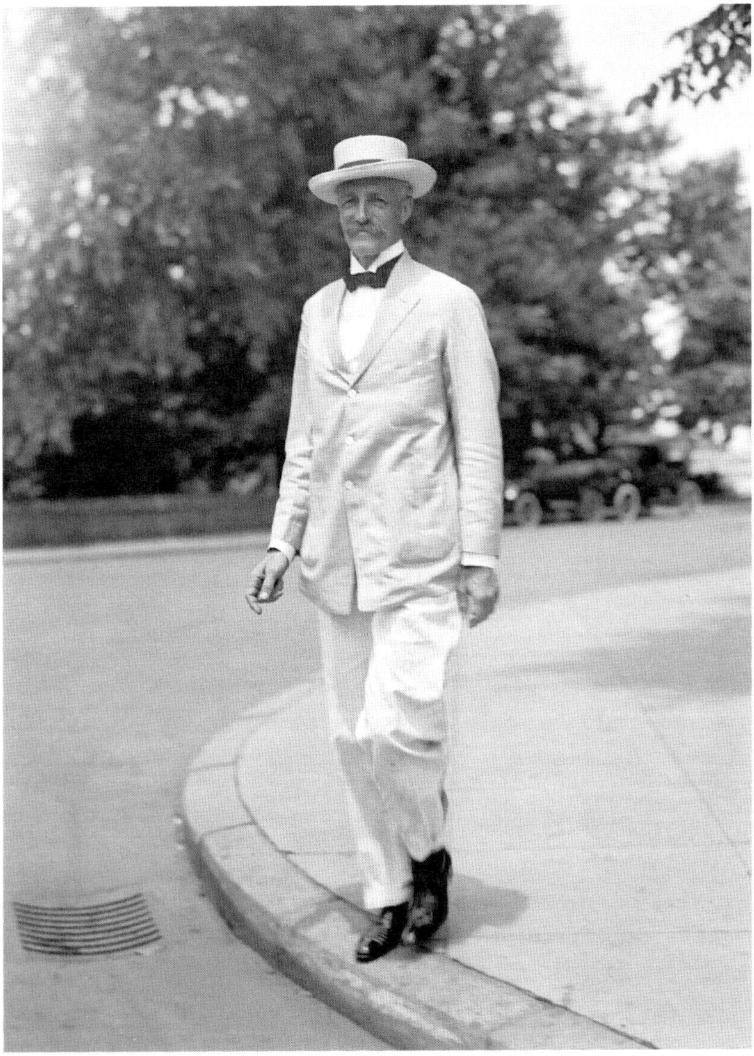

Gifford Pinchot was the first person to lead the U.S. Forest Service. He is widely considered the "Father of Conservation." *National Archives.*

But long before state and federal governments, as well as the forest service, developed a network of fire lookouts across the country, private landowners realized the need to watch for fires and built their own lookouts and towers.

Before Pinchot became the face of conservation and his successors fought for total fire suppression, wildfire consumed millions of acres of forestland in America. In the Blue Ridge Mountains, where economic hardship prevailed and natural resources were nearly depleted by unregulated industry, wildfire practically became procedural to development.

According to anthropologist William Jurgelski, fires were caused by a variety of sources in the Blue Ridge, including Native Americans, farmers, lumber companies, campers and early settlers. Many accidental fires were set during that time, Jurgelski noted, and most were preventable. He said that farmers burning land for tobacco fields and early settlers hiking through the mountains were known to accidentally set large fires, which could burn entire mountainsides.

There were laws that prohibited or restricted individuals from setting fires, similar to laws today, but many codes were either disregarded or ignored, or settlers and farmers simply did not know about them. Enforcement of burn bans was nonexistent in the Blue Ridge.

The late nineteenth-century railroad industry was one of the leading causes of fire in the Blue Ridge. Starting in 1880, railroads began cutting through the mountains. And like in many other regions of the country, wood- and coal-burning locomotives pulled trains in the Blue Ridge. These fire-driven trains ignited flames throughout the mountains. Sparks and ashes from passing trains regularly triggered wildfires. Although railroad companies attempted to prevent fires by maintaining clear right-of-ways, fires continued to be ignited.

Ferdinand Silcox, one of Pinchot's successors, wrote about fire prevention prior to taking the reins of the U.S. Forest Service. In a 1910 report, *Fire Prevention and Control of the National Forests*, Silcox said, "Detailed reports of the Forest Service for 1909 placed the railroads first as being the most common cause of fires on the National Forests. Out of 3,138 fires reported, 1,186 were caused by locomotives, and their setting was due to three principal reasons—the use of coal as fuel, the lack of proper clearing of the right-of-way, and the non-use or misuse of spark arrester." Silcox also wrote about lightning, farming and logging, three additional sources of wildfires in the Blue Ridge.

The logging industry, which reached a peak in the southern Appalachian Mountains between 1900 and 1915, resulted in mountainsides cleared by nearly 90 percent. Silcox wrote that there were thirty-eight fires caused by logging operations in the national forests in 1909. He also said that sawmills were a cause of fires in the forests.

Professional foresters, including Silcox, believed that many of these early fires in the Blue Ridge needed to be stamped out, leading to a vast fire prevention, detection and suppression network. In his 1910 report, Silcox said that ground patrol units and fire lookouts were necessary to detect fires:

> *After dividing the forest into ranger districts and subdividing the districts into patrol units, with a man in charge of each, it is still necessary to make sure that these men are in a position to render effective service. The ranger must select certain lookout points and ridges from which he can see over his entire district...A view from these high points will, in many cases, be worth a great deal more for discovering fires than patrol lower down; hence these points are carefully selected and coordinated to give primary control of the entire forest. They are generally high isolated peaks from which an unobstructed view may be obtained. If possible they should be in sight of each other, so that two men can locate a fire accurately by taking triangulation compass bearings. Yet these lookout points and ridges are of little value if after the fire is discovered there is no way to get to it quickly, because of a lack of trails, or no way to call for immediate help.*

Silcox additionally wrote about the need for developed trails and communications for lookout and firefighting operations.

The development of the fire lookout system became more organized in 1912, with Daniel Adams's *Methods and Apparatus for the Prevention and Control of Forest Fires, as Exemplified on the Arkansas National Forest.* Adams, who believed that a simple foot patrol in the nation's forests was inefficient, developed a network of towers in Arkansas. He proposed a system of "lookout and signal stations on the highest peaks, 10 to 15 miles apart, to locate incipient fire definitely, no matter where they might be."

Adams's proposal was considered a success. Adams, who joined the forest

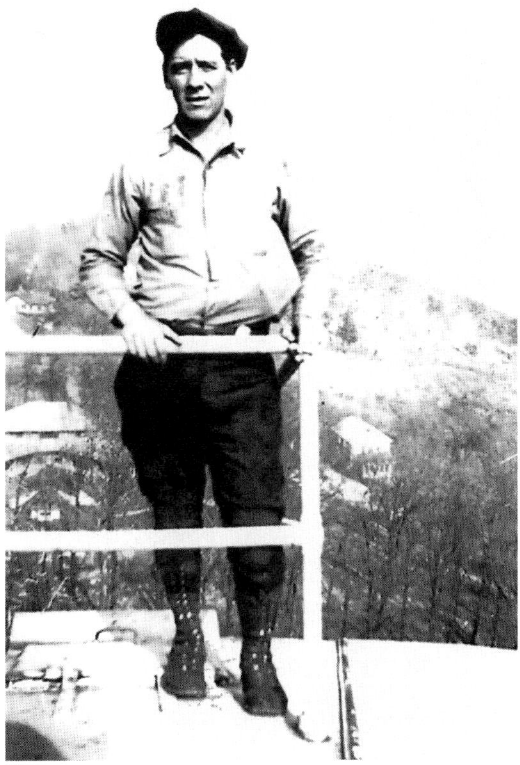

Betty Hodge's father traveled from Bristol, Tennessee, to assist in the construction of fire towers in the Blue Ridge Mountains. *Betty Hodge.*

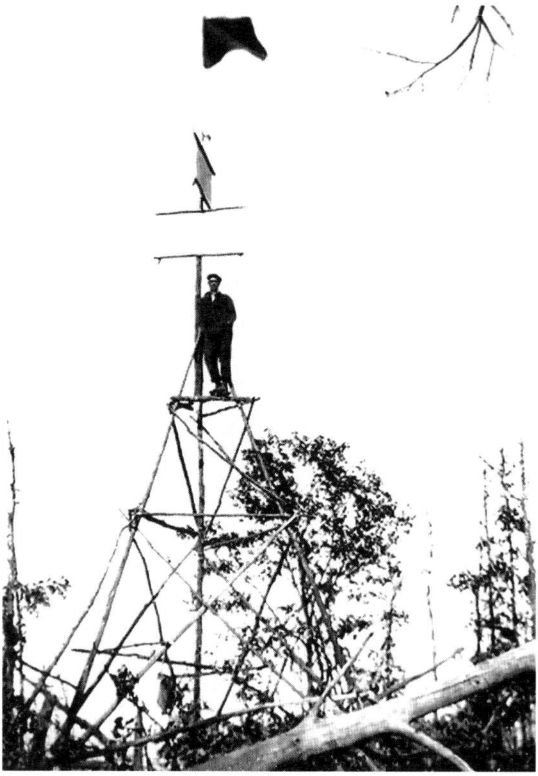

During the early years of fire lookout construction, men built makeshift towers of wood. This tower was located near Mount Pisgah. *Betty Hodge.*

service in Ashe County, North Carolina, designed lookouts, roads and trails, communication systems and equipment. He also developed a fire finder. Once the fire towers and lookouts were in place, fire damage on the Arkansas National Forest declined significantly. His system was replicated throughout the country, including North Carolina, to which he returned in 1914.

Prior to 1930, dozens of lookouts were established in the Appalachian Mountains by farmers, land speculators and coal companies.

During the early years of fire detection, land managers used a variety of means to watch for smoke and fire. Some landowners built wood cabins and huts. One- and two-story wood cabins were constructed on top of mountain balds, in fields and on rocky surfaces. The cabins included living quarters and spaces to view the terrain. Some cabins featured cupola roofs, where lookouts could watch for fires. These types of lookouts generally only worked at the highest elevations, unlike tall towers, which could be constructed at lower elevations. Large windows covered the exteriors of the cabins, so lookouts could watch for fires.

One of the earliest lookouts in the Blue Ridge, a cabin on top of Satulah Mountain in Macon County, North Carolina, was built in the 1920s, according to the Forest Fire Lookout Association. But lookouts didn't watch for smoke from inside the cabin; they worked at a post outside. An alidade, a piece of equipment used in fire finding, was placed on the post. A steel tower eventually replaced the cabin and post, but it has also been dismantled.

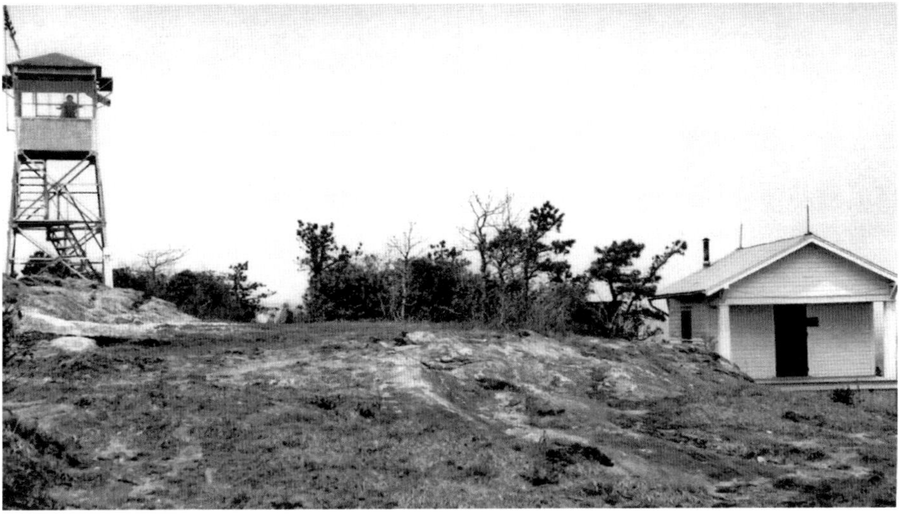

One of the first fire lookouts in North Carolina was located at Satulah Mountain. It was a one-story lookout cabin. A tower was later built. *National Archives.*

Alidades were also mounted onto trees and ladders. Tree lookouts were more common in the western United States and Australia, but Ron Stafford of the Forest Fire Lookout Association said there were likely some tree lookouts in the eastern United States as well. The South Carolina Forestry Commission reported that tree lookouts were used in the state's Lowcountry but noted that it was possible that trees were also used in the Upstate mountains. The state does not maintain records regarding tree lookouts.

Tree lookouts featured ladders or pegs that allowed individuals to climb the tree to reach a platform or chair. The platform was usually constructed on top of a stumped or lopped-off tree. Sometimes, as much as ten to fifteen feet was occasionally cut from the top of the lookout trees to accommodate a platform. An alidade may have also been mounted and attached to the lookout. Trees used for long periods of time were also equipped with telephone lines so the lookout could contact fire dispatch.

There were also many early lookouts that used no structure at all. Horse-mounted lookouts and individuals on foot have watched for fires since the colonial era. They would simply man the highest ridges, which were usually barren, and watch for smoke and flames. Mules, donkeys and oxen have also been used in fire detection. Lookouts have also been known to climb up on water tanks along railroads, utility poles and an assortment of other buildings and structures.

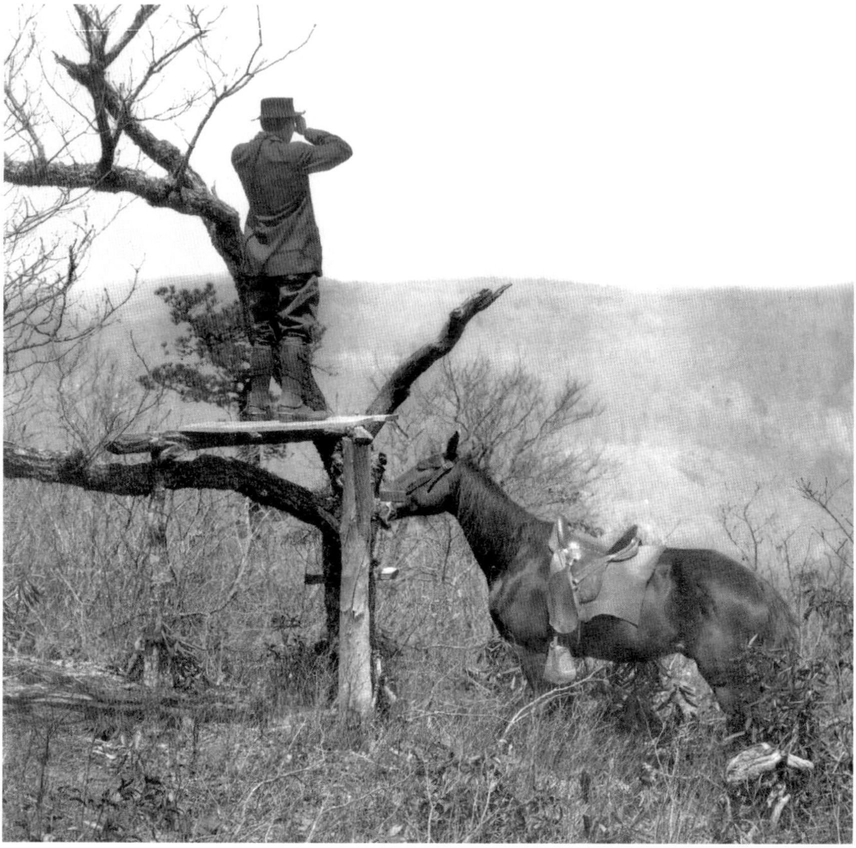

A man, who rode to the top of the mountain on his horse, stands on a small platform to watch for smoke. *Tennessee State Library and Archives.*

One of the first fire towers built in the Blue Ridge Mountains was located in Wise County, Virginia. The Virginia Coal Company, in association with the Virginia Department of Forestry, built the Stonega fire tower in about 1917. According to the state, it is likely the first lookout tower built in Virginia and perhaps in the Blue Ridge Mountains. Very little is known about the Stonega tower, which was located near Big Stone Gap in Wise County. The Stonega coal town, however, was established in 1896 and became one of the largest coal mining camps in the country. In order to watch for fires in the coal fields, the company built a tower and a small cabin on Little Stone Mountain.

A joint House resolution in Virginia stated, "The first fire tower was constructed on Little Stone Mountain near Big Stone Gap in 1917; that year, 1,460 forest fires burned 305,000 acres in Virginia."

It's unknown how long the fire tower was in use at Stonega or whether it was manned by the company or state or federal government. Through a succession of owners, the coal camp operated successfully until the mid-twentieth century. Although still listed on topographic maps in 2014, the Stonega Fire Tower has long been removed from the site.

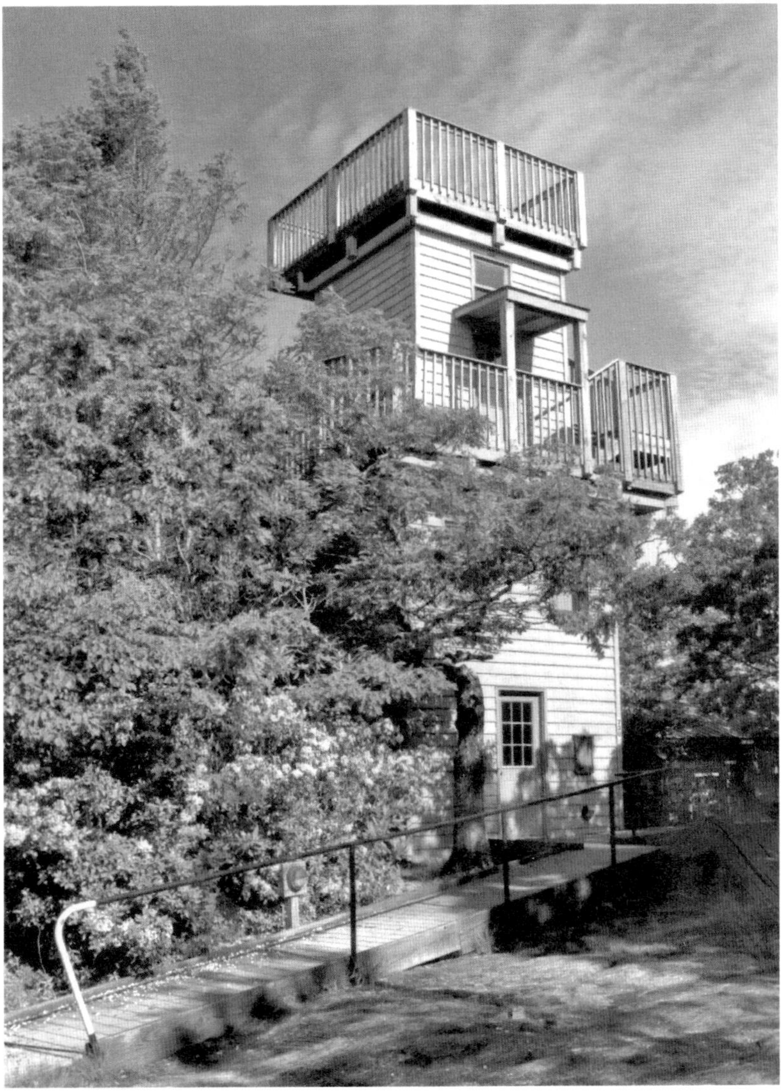

The old Camp Parrydise fire tower on Little Scaly Mountain is now used as an observation and entertainment venue. *Mountain Retreat and Learning Centers.*

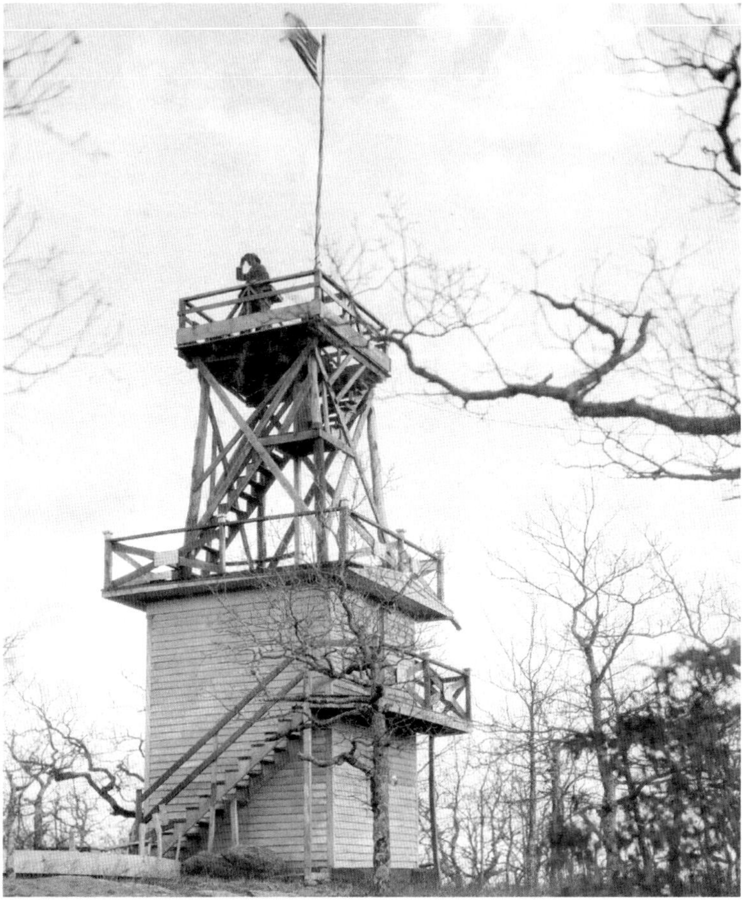

The Parry family built a fire lookout tower on Little Scaly Mountain in North Carolina to protect their all-girls summer camp. *National Archives.*

Coal and mining companies regularly built fire towers during the early twentieth century, especially in the Appalachian Mountains. The companies, which owned vast amounts of land in the Blue Ridge, sought to protect their investments and coal mining operations.

Another early lookout tower, known as Little Scaly Mountain or Parrydise, was built at an all-girls camp in the Blue Ridge of southwestern North Carolina. The Parry family of Decatur, Georgia, constructed a multi-story wood fire tower at their all-girls camp in Highlands around 1920. Judge Harvey L. Parry's family established the camp, and his wife managed it. The fire lookout building is located on the Eastern Continental Divide and was in good condition in early 2014. It has been restored and features multiple observation decks.

Ted D. Wisniewski, interim director of the Mountain Retreat and Learning Centers, a Camp Parrydise successor, said that the tower is located on one of the oldest fire tower sites on the East Coast. According to the Forest History Society, the Parrydise tower was owned by the camp but used by the forest service.

In recent years, Wisniewski said, the camp has used the tower to mount webcams and a weather camera. Guests also use it for viewing sunsets and sunrises, stargazing and hanging out. There are two small meeting rooms in the tower, and visitors can view four states from its decks: North Carolina, South Carolina, Georgia and Tennessee.

Wealthy landowner Moses Cone, who purchased 3,500 acres of property outside Blowing Rock, North Carolina, in the 1890s, built a lookout tower on Flat Top Mountain. Cone and his wife, Bertha, also built a beautiful twenty-room Colonial Revival luxury home on his estate in 1901. So the textile mill owner could see the entire estate, he constructed a chestnut log lookout tower on top of the mountain. He also constructed a carriage road to the vista.

After Bertha Cone died in 1947, the Blue Ridge Parkway and National Park Service acquired the estate. By that time, the wooden tower had deteriorated and was torn down for safety purposes. The government replaced Cone's lookout with a steel tower featuring an observation deck. The steel tower remains standing at Moses Cone Memorial Park.

One of Tennessee's earliest towers in the Blue Ridge Mountains was constructed near Elizabethton between Holston and Iron Mountains. The small lookout tower was built in the Stoney Creek community, which lies in a valley between the two mountains. Higher-elevation towers were also constructed on Holston and Iron Mountains, but the Stoney Creek tower was used to spot fires in the valley.

The tower was built on top of a barren hill overlooking Blevins Hollow. Elizabethton resident Phyllis Cline said that she remembers when Thomas Bishop, her father, briefly worked at the lookout. The lookout, she said, was able to see the entire Stoney Creek Valley, which covers several square miles. Cline provided a photograph of a young woman standing next to the lookout, which appears to be a white wooden building, likely a ground cab lookout.

The old Stoney Creek lookout was eventually dismantled and taken to a nearby residence on Ensor Graveyard Road, where it remained for a few years. The lookout site is now listed on topographic maps as the Stony Creek Fire Tower and has an elevation at 2,251 feet. There are a few remnants

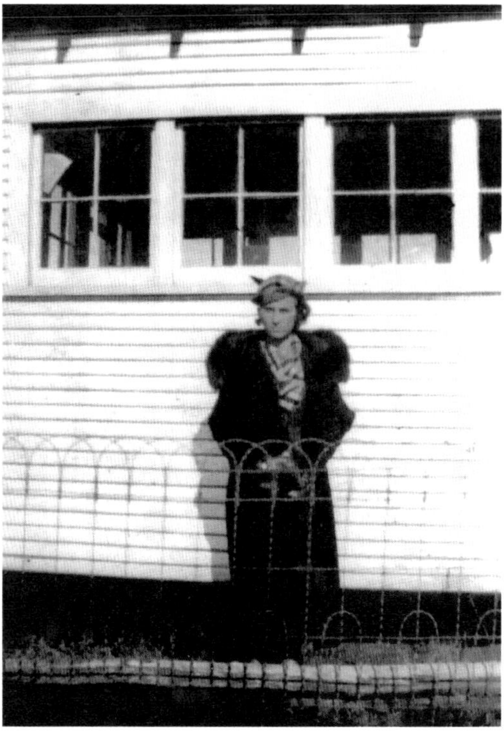

During the early and mid-1900s, a wood ground cab fire lookout stood on a short hill in the Stoney Creek community of Tennessee. *Phyllis Cline.*

of the old lookout on the hill, which is now covered in woods and a potato field, with glass and metal pieces embedded in the dirt.

According to the Tennessee State Library and Archives, the state's first fire tower was built on Cagle Mountain in Sequatchie County, near Chattanooga. The original wooden tower was eventually replaced with a steel tower, according to various sources. Few details are available of the old Cagle Fire Tower.

Conservationists Theodore Roosevelt and Gifford Pinchot led the government's efforts to protect the forests from fire and other crisis. Through Pinchot's leadership, in 1908, Congress enacted the Forest Fires Emergency Act, which authorized the forest service to spend whatever was necessary to combat the increasing wildfires throughout the country. Fire prevention and control programs immediately began to grow from these funds.

Two years after the Forest Fires Emergency Act was passed, a catastrophic outbreak of wildfires in the Northwest sparked further interest in firefighting and protecting the nation's forests.

By 1910, entire mountains throughout the country, including in the Blue Ridge, had been excavated for natural resources, including coal and lumber. Entire mountainsides were cleared of hardwoods and softwoods. The logging boom drastically changed the landscape of the Blue Ridge between the 1880s and 1920s.

The situation was similar in the Northwest, where the Great Fire of 1910—also known as the Big Blowup, the Big Burn and the Devil's Broom Fire—burned about 3 million acres in northeastern Washington, northern Idaho and western Montana. The fire, which was about the size of Connecticut, burned for more than two days and killed eighty-seven people. Most of the fatalities were firefighters trying to deter the flames. The Great Fire is believed to be the largest fire recorded in U.S. history. It was also the deadliest blaze for firefighters until the September 11 terrorist attacks in 2001.

Fire season started earlier than normal in 1910, according to U.S. Forest Service reports. The summer of 1910 was also hotter and dryer than usual. The extreme drought resulted in dry vegetation, which fueled the Great Fire. Although the exact cause of the fire is unknown, smaller blazes in the region were ignited from cinders flung from locomotives, lightning, sparks and backfiring crews.

By August 20, 1910, hurricane-force winds had whipped the smaller fires into one or two large infernos. The forest service, which had originated a few years earlier in 1905, was still small and had too few men and supplies to fight the blaze.

Due to the massive inferno in the Northwest, the army was brought in to assist forest service firefighters. Crews finally got the blaze under control when a cold front swept through the region, bringing steady rainfall. The rain came too late for many residents, however, as the blaze destroyed several towns, including Wallace, Idaho; Grand Forks, North Dakota; and Henderson, Montana.

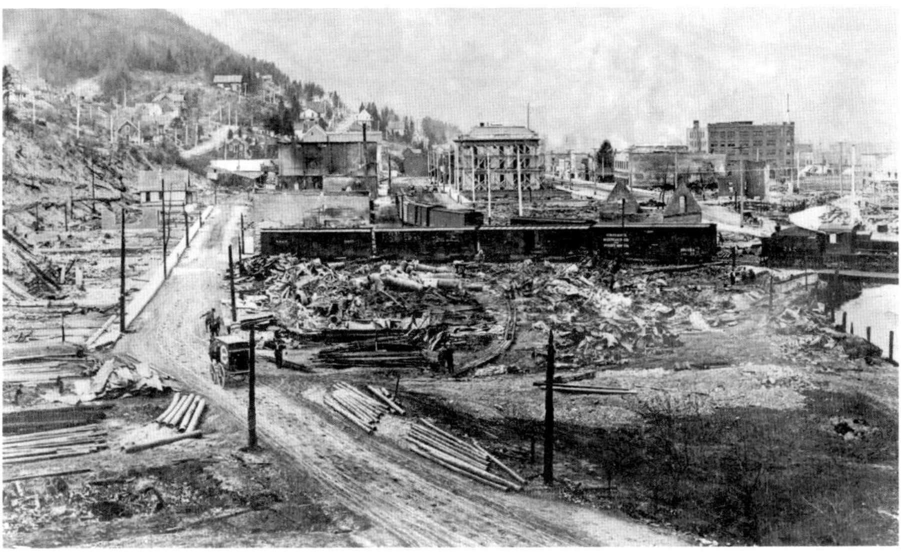

Massive fires destroyed several towns, including Wallace, Idaho, during the Great Fires of 1910. The disaster sparked a movement to protect the nation's forests. *National Archives.*

The devastating fire, which made news headlines around the world, triggered a whirlwind of interest in fire protection. One change was that the U.S. Forest Service required townships, corporations and states to bear the cost of contracting fire suppression services.

As a result, early fire detection became an issue across the country. And Pinchot's theory that some fires were healthy was largely ignored. After 1910, all fires were suppressed, no matter the cost. Fire detection networks developed in several areas.

Firefighting efforts developed in Maine, New Hampshire and Vermont, where fire tower networks were created. W.R. Brown, founder of the Brown Paper Company, helped establish the New Hampshire Timberland Owners Association. The organization is believed to have developed one of the first fire lookout networks in the country.

Smoke from the great Northwest fires literally and figuratively reached the nation's capital. In terms of response from Washington, D.C., fire protection came by means of legislation. In 1911, President William Howard Taft signed the Weeks Act. The Weeks Act, also known as the Appalachian Bill, authorized the federal government to purchase private land in order to protect the headwaters of rivers and watersheds in the eastern United States. It also called for fire protection efforts through federal, state and private cooperation. The federal government has subsequently purchased nearly 20 million acres of forestland from coast to coast, according to the forest service.

The controversial conservation bill actually took more than a decade to pass in both the U.S. House of Representatives and the Senate. Fiscal conservatives said that they did not want to put money into the nation's forests. Political conservatives said that the federal government didn't have the authority to purchase private lands. They even wanted the government's forestry department abolished. Others said that states should control the forests, not the federal government. The 1910 fires may have tipped the scale and stoked the bill's passing.

The National Forest Reservation Commission was created by the Weeks Act. The commission was used to consider and approve the purchase of land. On March 7, 1911, the commission held its first meeting. Less than a month later, forest service chief Henry Graves submitted recommendations for land to be purchased in thirteen areas of nine eastern U.S. states. Only eleven properties were purchased. The first purchase made under the Weeks Act was in McDowell County, North Carolina, which became part of the Pisgah National Forest. Land at Mount Mitchell, the location of a fire lookout, was one of the first purchases.

The Weeks Act also provided funds for fire watchmen's salaries. The act set aside $200,000 in matching funds to be distributed to states for forest protection. The agencies could also apply for $10,000 to be used for salaries if states would match the amount. In addition, the act encouraged states to expand state forests.

By 1920, the government had purchased more than 2 million acres of land under the Weeks Act. The Clarke-McNary Act of 1924 expanded the earlier Weeks Act. As a result of the new act, the government could purchase any lands that were once in timber or that could be used to produce timber.

By 1929, eight national forests had been established in the southern Appalachian Mountains covering about 2 million acres. The Blue Ridge region was considered the most significant area to be affected by the Weeks Act and the Clarke-McNary Act.

There were some issues resulting from the purchase of forestland, including an increase of "revenge fires." Many residents, who were already suffering from the economy, were angered by the fact that the government had come in and bought their lands. Several fires were set by these angry residents. Even Biltmore Estate, which owned thousands of acres of forestland, was targeted by arsonists.

Less than a decade later, and more than twenty years after the big Northwest fires, a large number of fire lookouts were finally constructed across the country to protect the nation's forests. It took not a natural disaster but rather the most devastating economic disaster in American history to construct fire lookouts across the country.

The Blue Ridge became ground zero for lookout construction during and after the Great Depression, which not only changed the way of life for millions of Americans but also reinvigorated its forests, bringing in thousands of workers and money to protect wildlands.

The world's stock markets had begun to decline in late September 1929, and on October 29, 1929, commonly known as Black Tuesday, Wall Street crashed. Black Tuesday pushed the country to the brink of an economic collapse. Although a true economic collapse did not happen, the American economy faltered for the next few years, and millions of residents lost their jobs. By 1933, at the height of the Great Depression, about 25 percent of the nation's population was unemployed. Wages had dropped by about 42 percent. Businesses shuttered. Farms halted production and closed. Young men who were starting families became unemployed across the country.

In 1932, Franklin Delano Roosevelt was elected president on his promise of creating new federal programs that he said would jumpstart the economy. He was inaugurated in March 1933, and over the next one hundred days, Roosevelt and his advisors developed the New Deal. In addition to the Banking Act of 1933, which the Roosevelt Institute noted halted economic panic in the country, a handful of new agencies were created. Key agencies included the Works Progress Administration (WPA), Civilian Conservation Corps (CCC) and the Agricultural Adjustment Administration. The Social Security Act and Securities and Exchange Commission were also created as a result of the New Deal.

The "Tree Army," as President Roosevelt's CCC became known, was launched on April 5, 1933, with the issuance of an executive order. The order appointed Robert Fechner as director of the new agency. In addition, the secretaries of war, agriculture, the interior and labor were informed to appoint representatives as advisors to the CCC's new director; $10 million was also set aside for the construction of public buildings, including fire towers and other forestry structures, by the new agency.

The president spoke about the CCC on May 7, 1933, during one of his famous "Fireside Chats," which laid out the administration's plans to curb the nation's banking crisis:

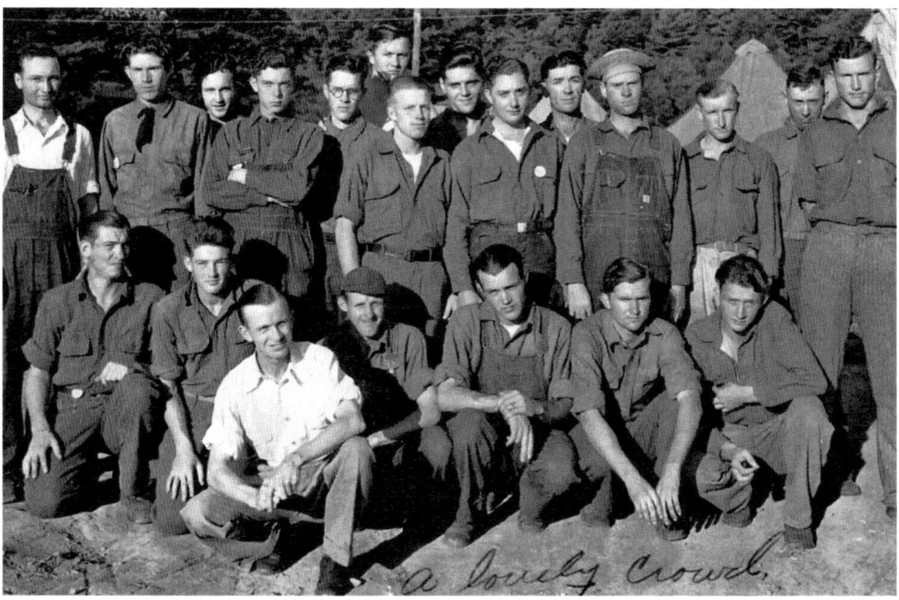

Men enrolled at the Civilian Conservation Camp in Hampton, Tennessee, worked on a variety of projects, including the construction of fire towers. *Phyllis Cline.*

First, we are giving opportunity of employment to one-quarter of a million of the unemployed, especially the young men who have dependents, to go into the forestry and flood prevention work. This is a big task because it means feeding, clothing and caring for nearly twice as many men as we have in the regular army itself. In creating this civilian conservation corps we are killing two birds with one stone. We are clearly enhancing the value of our natural resources and second, we are relieving an appreciable amount of actual distress. This great group of men have entered upon their work on a purely voluntary basis, no military training is involved and we are conserving not only our natural resources but our human resources. One of the great values to this work is the fact that it is direct and requires the intervention of very little machinery.

Thousands of men seeking employment joined the CCC, which completed many tasks, including the construction of fire towers. Originally, the agency required enrollees to be between the ages of eighteen and twenty-three. Eventually, however, it expanded to young men ages seventeen to twenty-eight. The agency's maximum enrollment at any time was 300,000. Between 1933 and 1942, when the agency ceased operation, about 3 million men participated. Enrollees were given a small wage of thirty dollars per month, of which twenty-five dollars had to be sent home to their families.

Reserve officers from the U.S. Army were in charge of the CCC camps, which were established across the country. Quite a few camps were established in the Appalachian Mountains. On April 17, 1933, the first camp, NF-1, known as Camp Roosevelt, was established in the George Washington National Forest near Luray, Virginia, in the Shenandoah Mountains.

According to an article in the *Johnson City Press* in 2013, Kayla Carter reported that the CCC managed "practically all the fires in the area," spanning about "40 miles by 100 miles," with six lookout towers. Carter quoted a CCC pictorial review pamphlet provided by Heritage Museum curator Martha Erwin.

Unicoi County, Tennessee native Pete Chitwood told Carter about his time serving at Camp Cordell Hull in Company 1472. He noted that Unicoi's Pinnacle fire tower, which was restored in 2011, was used by the CCC as a fire lookout. In addition to building and manning fire towers, Camp Cordell Hull worked on a variety of recreation areas and trails in Unicoi County. Chitwood told Carter that he is proud of the work the local CCC camp accomplished because they helped to improve the area for future generations.

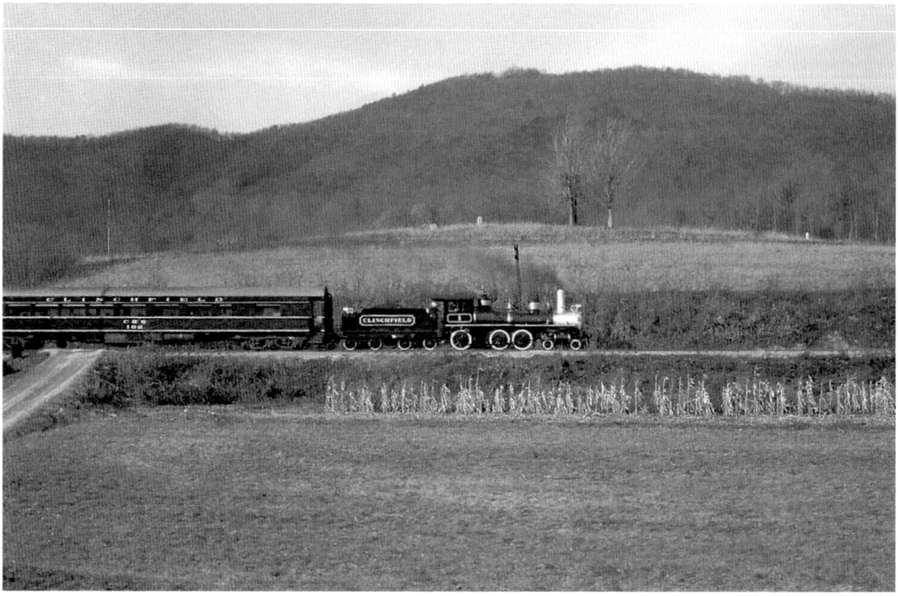

The Clinchfield Railroad rolls through the valley in Unicoi County, Tennessee, below the historic Pinnacle Mountain Fire Tower. *Mark Stevens.*

Throughout its inception, men in the CCC worked on a variety of projects, many of them aimed at protecting the forests. And most of the organization's projects involved the National Forest Service and National Park Service. In the Great Smoky Mountains National Park, CCC workers built roads, park structures and fire lookout towers.

Frank Davis, author of *My CCC Days, Memories of the Civilian Conservation Corps*, recalled joining a camp near Bryson City, North Carolina, when he was eighteen years old. In the book, Davis said that he was assigned to the trail crew. He later became a camp mechanic. "Our job was to build fire trails throughout the mountains to give access for firefighters to combat forest fires, and to build hiking trails for recreation use," he wrote.

Although Davis does not discuss fire towers in his book, members of the camp constructed and manned a few of the lookouts in the Great Smoky Mountains National Park. According to the federal government, the CCC built about five thousand fire towers across the country by the late 1930s. In addition to building newly established towers, the CCC also replaced some of the older wooden fire towers.

A CCC camp in north Georgia assisted the local forester in building a new replacement tower on Brasstown Bald, the state's highest peak.

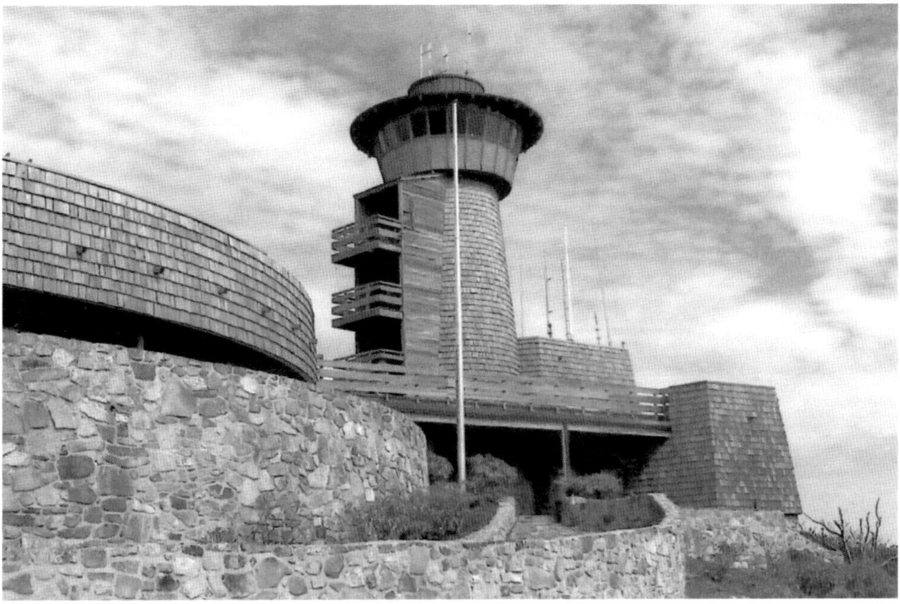

The Brasstown Bald Observation Tower was built in the 1960s to replace an older steel fire tower. *U.S. Forest Service.*

According to the Union County Historical Society, the first tower on Brasstown Bald, also known as Enotah Bald, was built in the 1920s by the Pfister-Vogel Logging Company of Milwaukee, Wisconsin. It was constructed of chestnut and locust wood and was built so that watchers could see smoke from any forest fires within the areas where the lumber company was conducting logging operations.

On thousands of acres the Vogels owned in northern Georgia, the family harvested bark from oak and hemlock trees, shipping it back to Wisconsin for tanning leather. During World War I, a synthetic method to tan leather was developed, so the company no longer needed Georgia's natural resources. The Vogel family gave its land to Georgia to create a state park. During the family's time in Georgia, most of the company's land was never cut.

Brasstown Bald stands in the Chattahoochee-Oconee National Forest. Between 1911 and 1930, the government bought about 743,000 acres of land and set it aside as a preserve.

Future forest ranger Arthur Woody was born in 1884 in a small log cabin in Suches, Georgia, a short distance from Brasstown Bald. According to legend, Woody witnessed his father kill the last surviving whitetail deer in Georgia, at a time when laws did not dictate hunting in the state. He decided at a young age that he would restore the deer population in the state and protect the forests.

In 1912, Woody began his career with the forest service as an axeman on a baseline crew. As he cut through the brush for other crewmembers, he learned about forest fire control and prevention. He also became a surveyor of lands for the forest service.

A few years later, Woody was sworn in as a forest guard, a job that required him to protect the forests from fire, trespassers and poachers. In 1918, he became one of the first forest rangers in Georgia. He oversaw the Blue Ridge District, which later became the Chattahoochee National Forest. As a forest official, he worked hard to encourage the government to purchase forestlands, assisted in maintaining the deer population and introduced fish into the streams.

In 1935, Woody worked with the CCC to build a new fire tower on Brasstown Bald to replace the earlier lookout built by the Pfister-Vogel Company. Camp Woody, a CCC camp, was established at the ranger's hometown in Suches. The CCC workers assisted Woody with a variety of projects, including the planting of trees and checking tree blights and insect infestations; fought forest fires; and built roads and Vogel State Park, which was established as one of Georgia's first state parks in 1931. Land was donated by the Vogel family for the state park. The CCC also built fire towers on the north Georgia lands, including one on Brasstown Bald.

According to the information provided by the historical society, Woody sat at his kitchen table and drew plans for the stone and wood tower. Men from Camp Woody followed an old logging trail up the mountain, where they set up camp and constructed the tower using local materials.

Woody's tower did not last long. After World War II, the stone and wood structure was replaced by a steel tower in 1947. Woody died in 1946, and according to several sources, including his obituary, he died of high blood pressure and heart attacks, apparently brought on by the stress of protecting his beloved forests. The steel tower was replaced in the 1960s with the most recent stone observation tower, which features an exhibit dedicated to Woody and the mountain's history.

Ranger Roscoe Nicholson, known by many as "Ranger Nick," grew up in Rabun County in north Georgia. In 1912, he became Georgia's first forest ranger. Nicholson and Woody are considered the fathers of forestry in Georgia and two of the most influential foresters in the nation.

Rabun Bald, located in Nicholson's community, is the second-highest point in Georgia, behind Brasstown Bald. According to the Rabun County

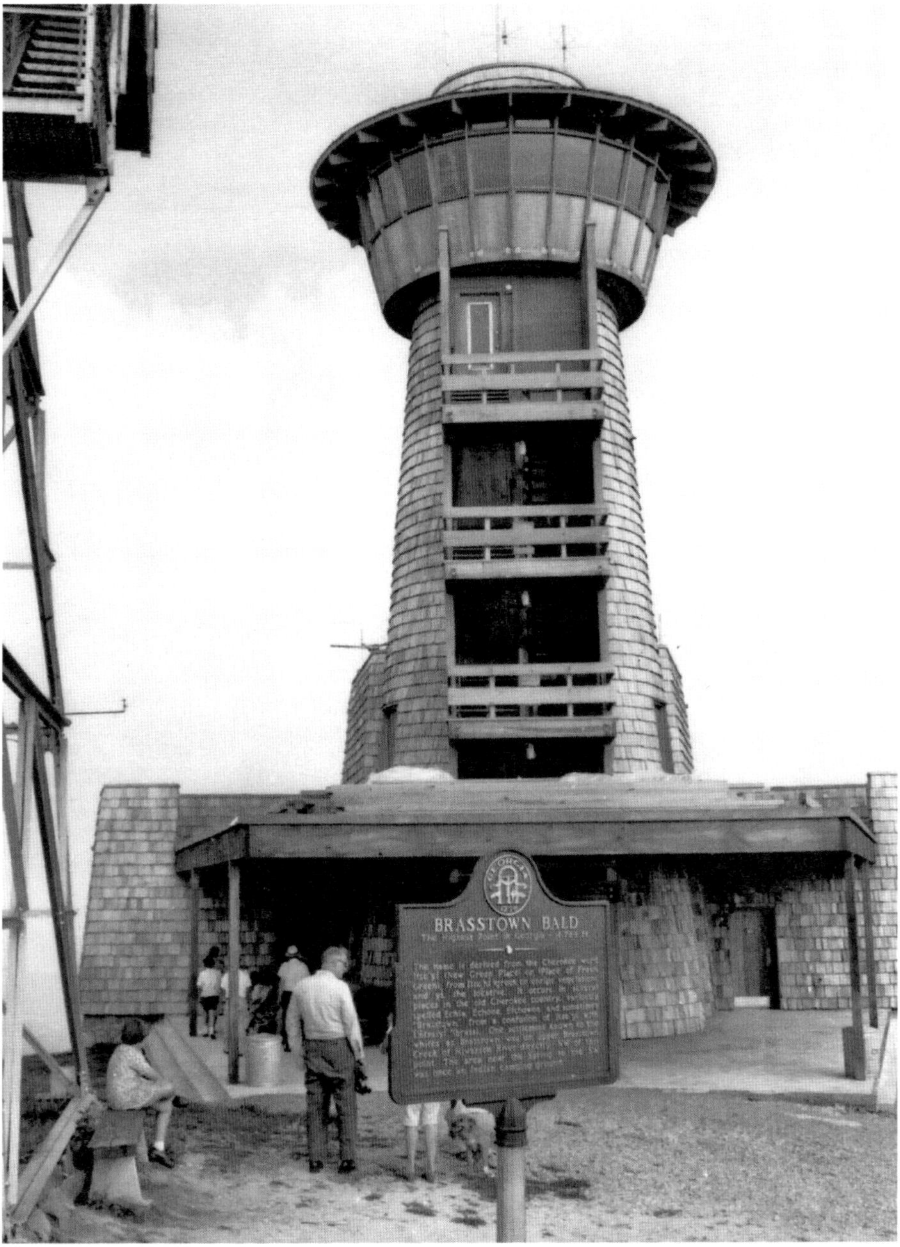

For a brief time in 1960s, the Brasstown Bald Observation Tower in north Georgia stood next to the old steel fire tower. *Union County Historical Society.*

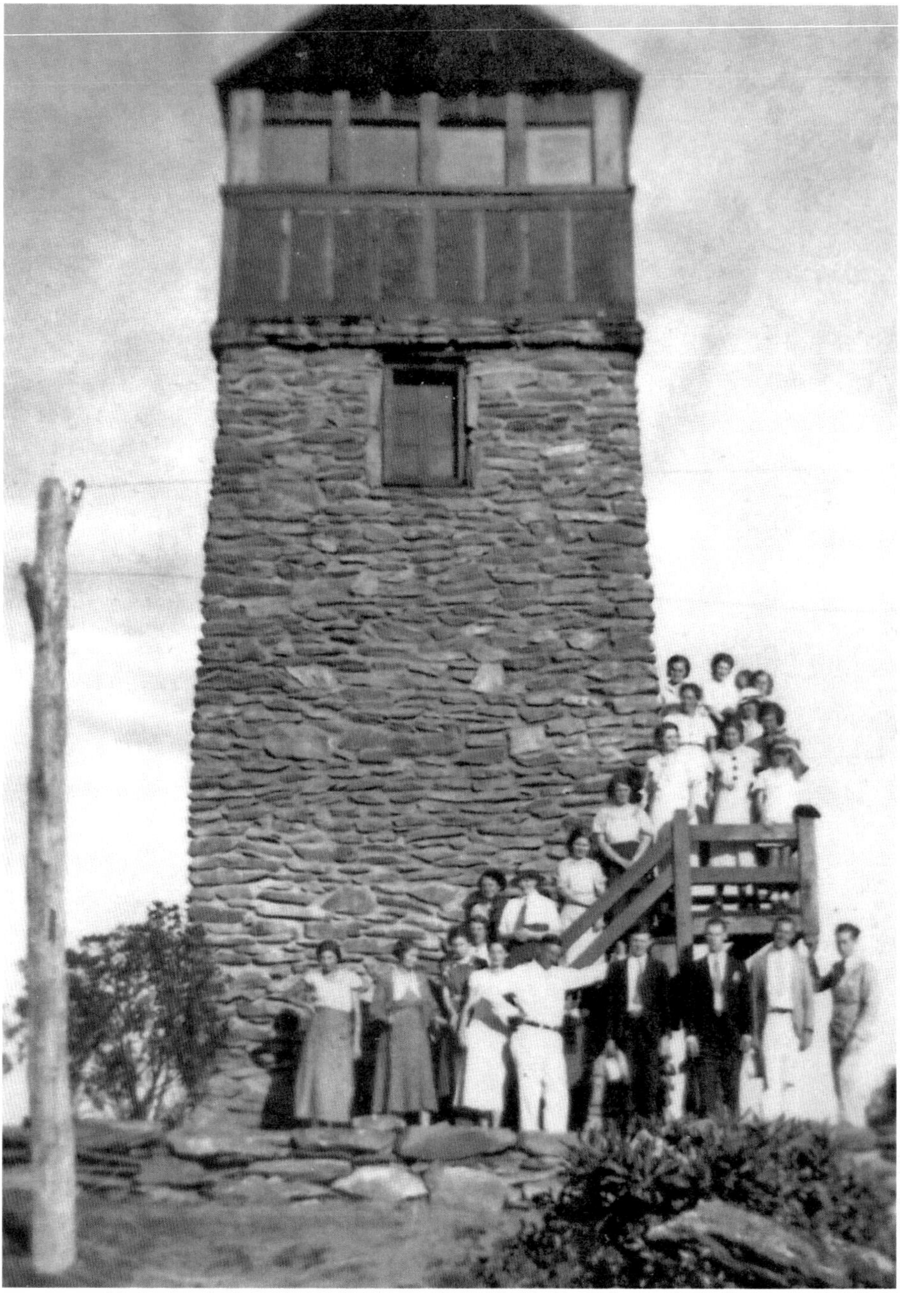

A group of visitors poses around the base of the stone tower that once stood on Brasstown Bald, Georgia's highest mountain. *Towns County Historical Society.*

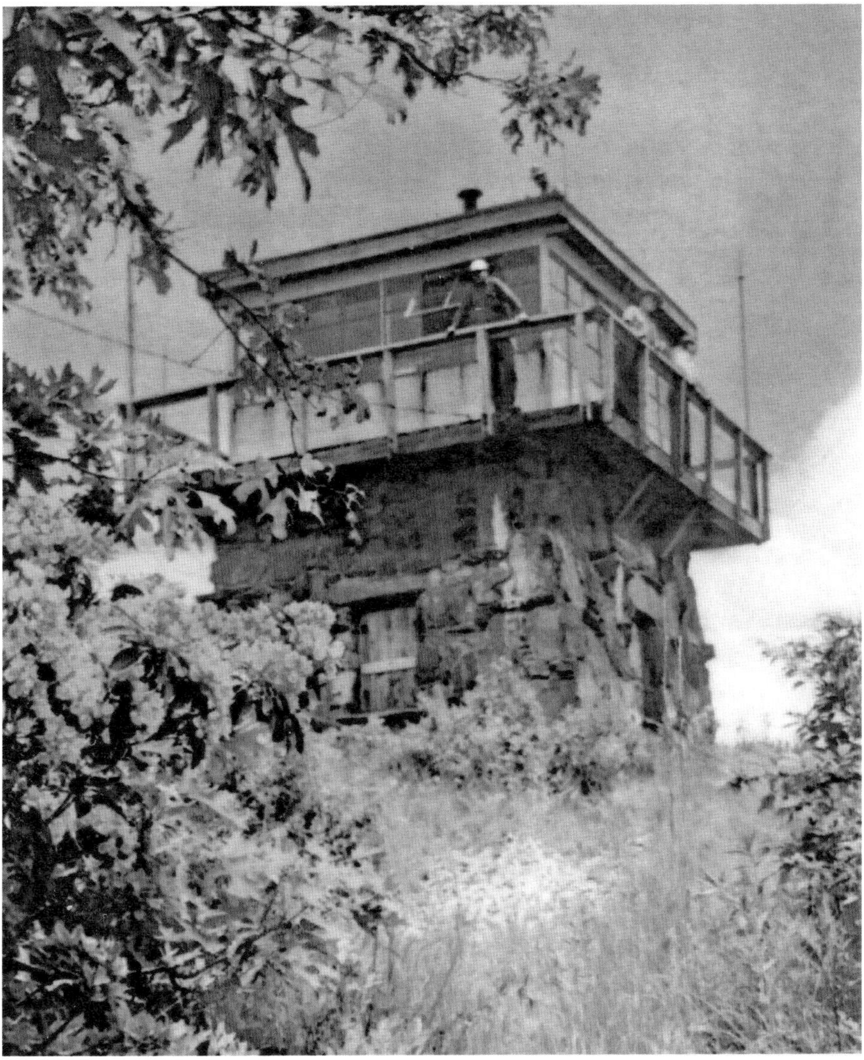

The Rabun Bald stone and wood fire lookout, built by the Civilian Conservation Corps, is located on Georgia's second-highest mountain. *Rabun County Historical Society.*

Historical Society, Nicholson assisted in the construction of a fire tower on the bald. Like Woody, Nicholson was concerned about forest fires in the area. In addition to constructing a fire tower, Nicholson is known for using bloodhounds to track down arsonists.

The stone lookout tower on Rabun Bald was constructed by workers from the nearby CCC Camp Warwoman in 1939, according to information provided by the historical society. Fire finding equipment was located in a glass room on top of the stone portion of the tower. The CCC men also installed telephone lines from tree to tree to reach the top of the bald. After about forty years of service, the lookout was decommissioned in the 1970s, and metal railing was installed on top of the stone structure for visitors. The fire lookouts on Brasstown Bald and Rabun Bald remain popular attractions in north Georgia.

The Little Snowball Tower, which now stands at the Big Ivy Historical Park near Barnardsville, North Carolina, was originally built in August 1934. The Big Ivy Historical Park happens to be located on the former site of a CCC camp, which built the twenty-one-foot-tall tower on nearby Little Snowball Mountain.

Barnardsville native John Mote Allen joined the CCC like many other rural men. As a member of the CCC, Allen assisted in the construction of the Little Snowball Tower. When World War II started, as other men joined the armed forces, Allen, who was a bit too old, began working for the forest service. He was hired to work on Little Snowball. The tower features a fourteen- by fourteen-foot live-in cab with a bed, stove and fire

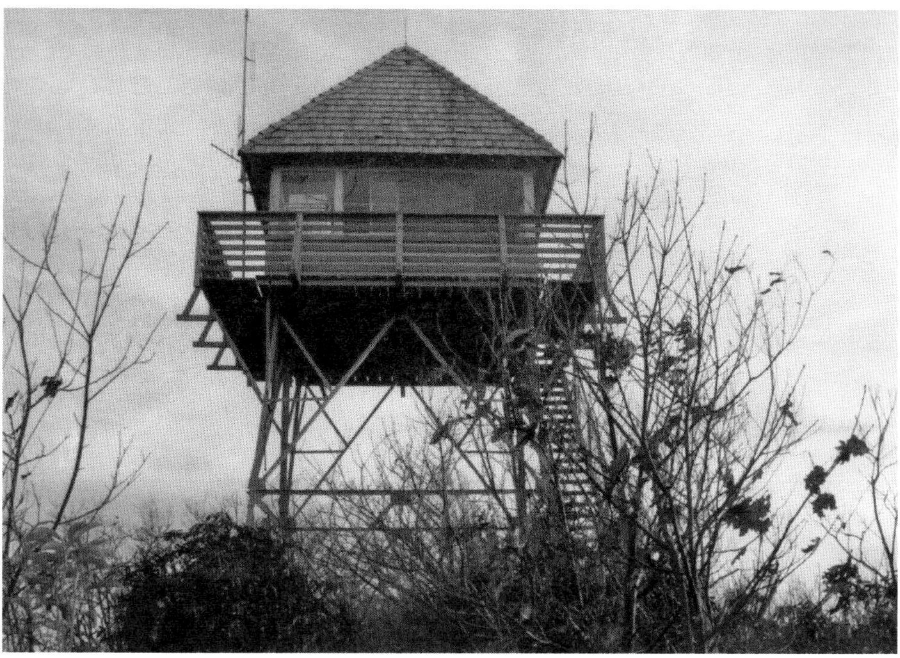

The Little Snowball Fire Tower no longer stands on top of the western North Carolina mountain. It has been rebuilt at a community park. *Lloyd Allen.*

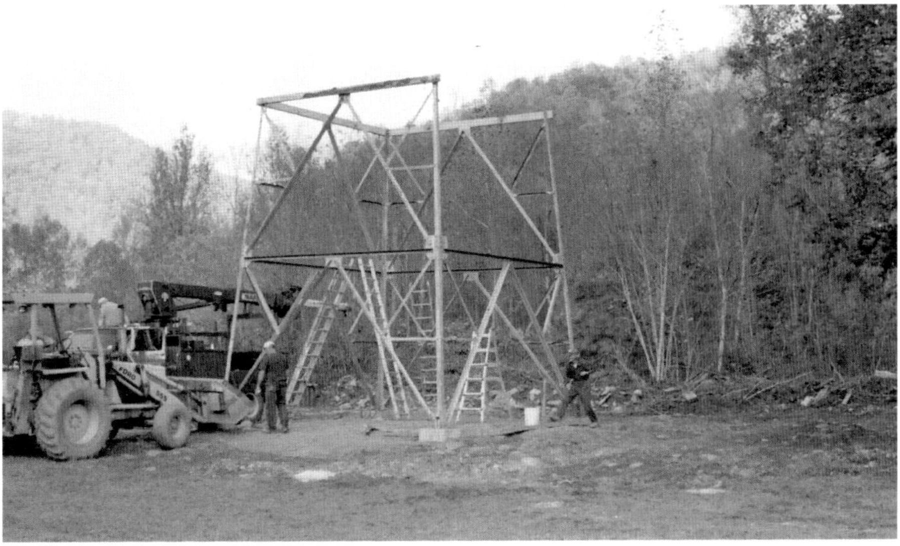

Lloyd Allen and his friends rebuilt the Little Snowball Fire Tower at the Big Ivy Historical Park. *Lloyd Allen.*

finding equipment. In the 1950s, Allen was replaced by Cling "Fabe" Webb, who worked on the tower for about twenty years, until the lookout was decommissioned.

Lloyd Allen said that in 1980, he learned that the forest service was interested in removing the tower. It had deteriorated due to conditions on Little Snowball, so the forest service wanted it removed for safety reasons. After talking to a forest service official, Lloyd Allen made an offer to purchase the historic tower for $300. He wasn't quite sure that the forest service would accept the offer, but it did. He bought the tower and, with help from his friend, Bill Hensley, dismantled it, piece by piece. They made sure to take every part, including all of the bolts and a piece of concrete used as a step. The tower was then stored in a small building in nearby Burnsville, North Carolina, until 2005, when Allen and his friends reconstructed the tower on the site of the old CCC camp at Big Ivy.

Allen, who remains a staunch supporter of fire tower preservation, said that his father-in-law, Blaine Ray, was the last fire lookout on nearby Green Knob Tower. Allen said that he regularly trekked up the mountain to visit his father-in-law, cook dinner and spend the night. The tower on Green Knob was constructed in 1931, two years before the CCC was established. Located near Mount Mitchell, the tower was decommissioned in the late 1970s.

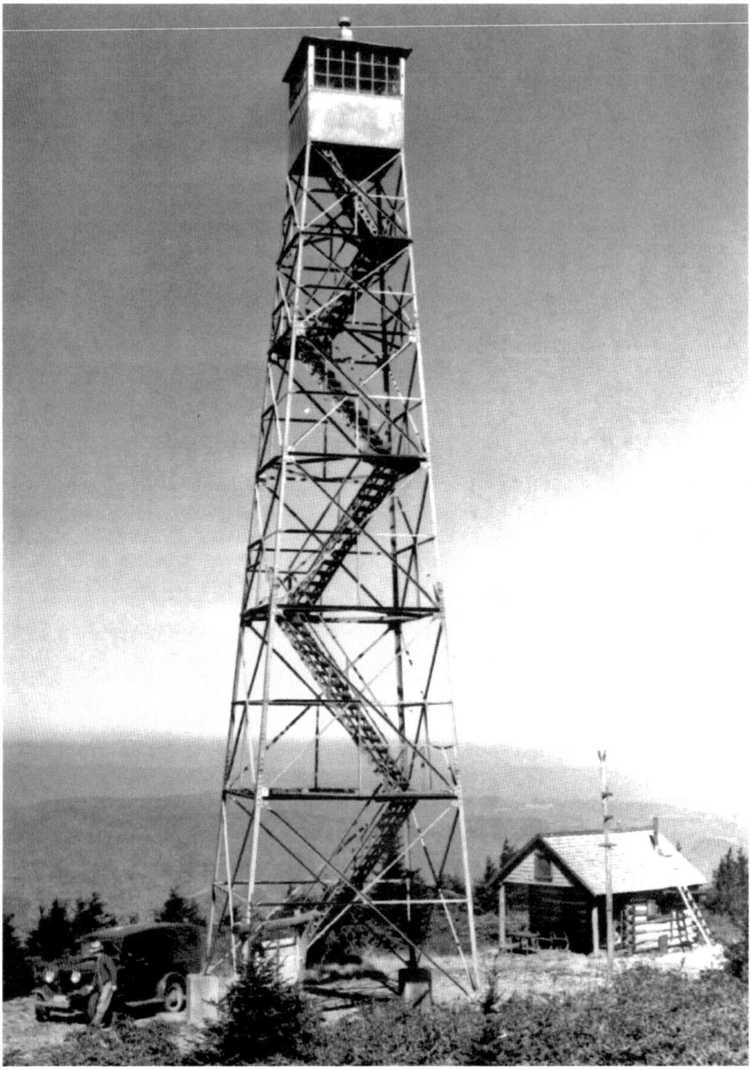

A fire tower stood on Roan High Knob, along the Tennessee and North Carolina border, for one decade. *Tennessee State Library and Archives.*

Across the state line in Tennessee, the CCC also built a tower on Roan Mountain. The mighty Roan, at 6,285 feet above sea level, was thought to be an ideal location to construct the highest fire lookout in the state. After all, on a clear day, one could potentially see five states. The tower was constructed in the mid-1930s. It was one of several CCC projects in Carter County, where Roan Mountain is located. Other projects included schools, park structures and roads.

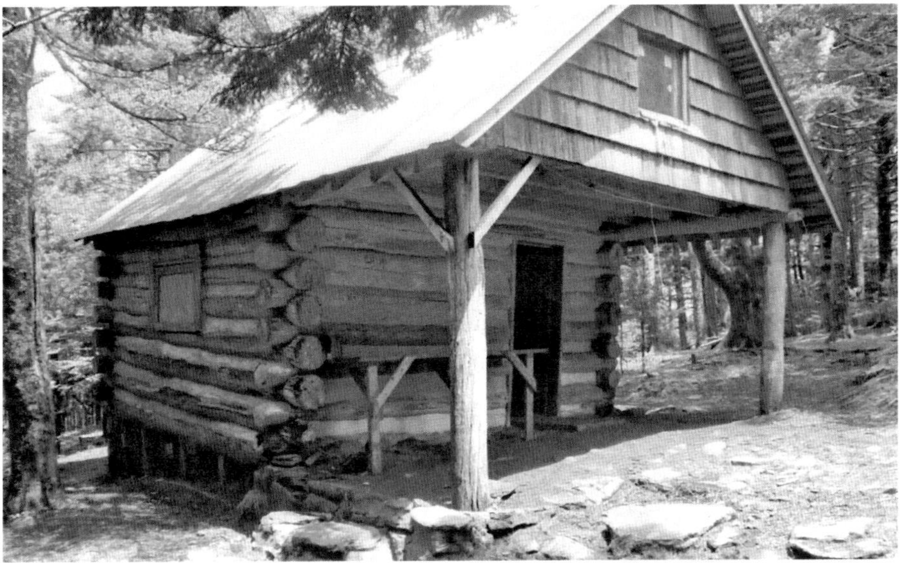

The Civilian Conservation Corps built a tower and cabin on Roan High Knob in the 1930s. Only the cabin remains standing. *Photo by Robert Sorrell.*

Although no records have been located regarding the CCC's construction of the tower, Camp Folsom Angel was established in 1933 in nearby Hampton, Tennessee. One of the camp's most important duties was to help fight and prevent forest fires. It also helped construct a few local roads.

Despite good intentions, the Roan High Knob Tower was rather useless. There weren't many trees on the Roan when the tower was constructed. Lumber companies had stripped the mountain of forests, leaving behind hundreds of acres of barren land. People, like East Tennessee State University professor Dr. Dalton M. Brown, had come to the Roan to study the tree life in the hopes of stopping the destruction of the lumber companies. Brown spent time on the mountain in the mid-1930s, at the time the tower was in use. He photographed the barren landscape many times, even capturing the tower a few times.

During its short-lived time span, the tower was a popular spot for visitors. The Dellinger family of Mitchell County, North Carolina, said they remembered visiting the Roan to picnic at the base of the tower.

Little is known about who managed the lookout, but it is known that it was used for only a few years. The Roan, much like the Great Smoky Mountains to the southwest, is often shrouded by low cloud cover and fog. That's why the mountain has been nicknamed "Cloudland." So, a majority of the time, the lookout, even watching from the tower's cab, couldn't see anything except for clouds. On a clear day, however, the 360-degree view from the tower's cab must have been incredible. The Roan's magnificent rhododendron gardens, barren balds and what was left of the spruce/fir forests, are quite a sight today; one could only imagine what they looked like from the fire tower.

Civil War general John T. Wilder even named his mountaintop hotel, which was located near the tower site, the Cloudland Hotel. The hotel existed from the late nineteenth century to the mid-1910s.

Although the tower was dismantled sometime around 1940, the fire lookout's cabin was kept intact. In fact, the log cabin, originally built by the CCC, still stands today on a spur of the Appalachian Trail. The one-and-a-half-story cabin, which is located near the base of the former tower, has been converted into a trail shelter. The Tennessee Eastman Hiking Club maintains the shelter.

According to author and historian Jennifer Bauer, the fire tower site is a popular spot for treasure hunters. In her book *Roan Mountain: History of an Appalachian Treasure*, she said that treasure hunters search for CCC remnants left at the site.

The Tennessee State Library and Archives reports that the tower was maintained by the state forest service. Photographs taken in 1938 reveal that telephone lines were installed running to the top of the mountain so the lookout could contact firefighting crews. In 1941, the federal government purchased large amounts of land on Roan Mountain to become part of the Cherokee and Pisgah National Forests.

A cabin, a wood tower and a stone structure have stood on Wayah Bald in the Nantahala National Forest as lookouts since 1927. The stone lookout, which still stands in 2014, took two years for the CCC to build, between 1935 and 1937, although the original three-story structure has since been shortened.

The U.S. Forest Service built the first lookout, a small twelve- by fourteen-foot wood cabin, on scenic Wayah Bald in 1927, according to a study by Edwards-Pitman Environmental Inc. Later, in 1929, a wood tower was constructed on the bald adjacent to the cabin. Once the CCC was established in 1933, the organization began improving the road up to the bald. In 1935, CCC workers began construction of a new sixty-foot-high native stone tower. It was a three-story structure, and the top level housed fire finding equipment. CCC enrollees hauled stone from local quarries for the new tower, according to the Edwards-Pitman study.

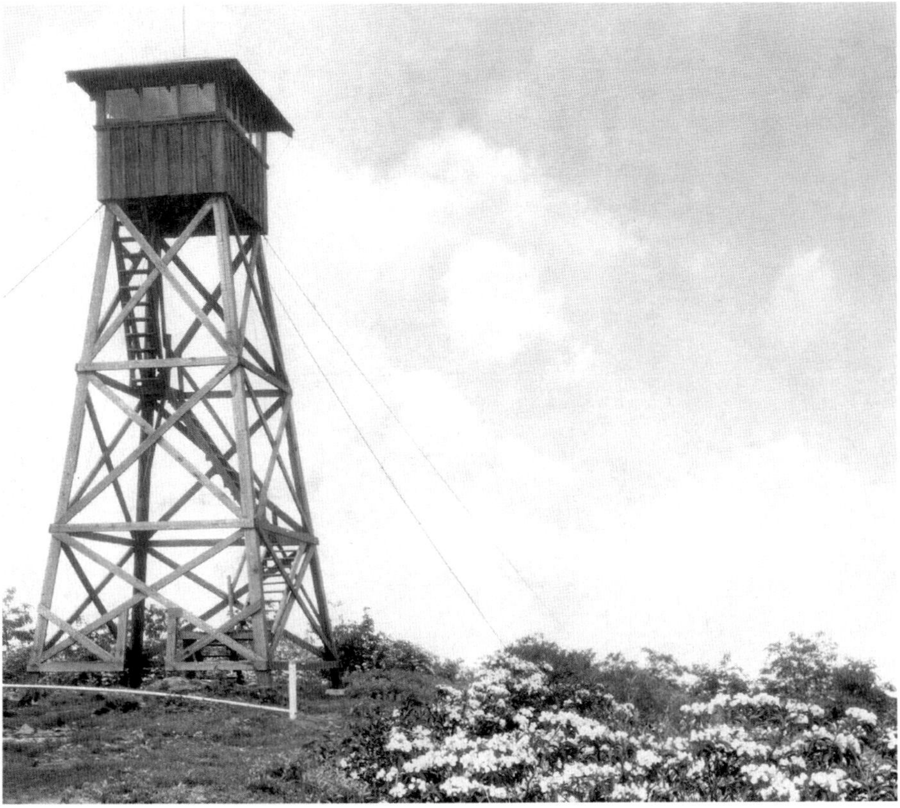

Before the government built a stone tower on Wayah Bald, there was a wooden lookout on the peak. *National Archives.*

In 1937, it was dedicated to former Nantahala National Forest supervisor John Byrne, who died in 1933. The tower, with an interior oak staircase, also featured a second-story public observation level with an exterior wooden catwalk. But by the mid-1940s, hairline cracks had begun to appear in the stone structure, resulting in damage, and the forest service decided to end lookout service at the stone tower.

After discontinuing use at the tower in 1945, the forest service decided to remove the top eighteen feet of stone below the cab and established an observation deck so visitors could still view the surrounding mountains. The dismantled stone was used to build an exterior stair on the east and north sides of the tower, the study reported. Located along the Appalachian Trail, Wayah Bald, with an elevation of 5,385 feet, remains one of the country's most popular and visited lookout sites. The forest service restored the tower in 1983.

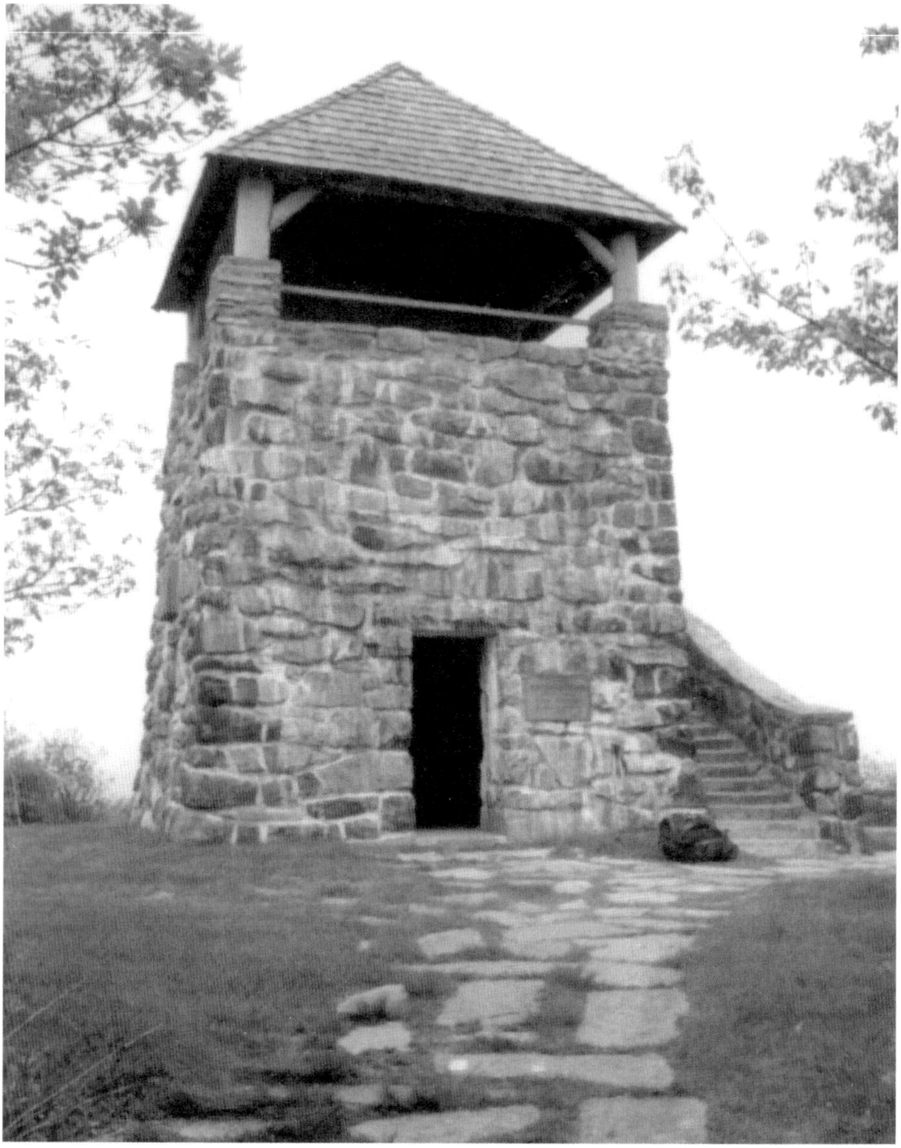

The stone lookout tower on Wayah Bald has become a popular tourist destination and is a regular stop for Appalachian Trail hikers. *Photo by Robert Sorrell.*

According to the forest service, the Wilson Lick Ranger Station, which is located along the gravel road leading up to the bald, was the first ranger station in the Nantahala National Forest. Located about three miles from the lookout tower, the ranger station was established in 1913 and used until

1967. From here, rangers maintained and managed the nearby fire lookout, as well as management of the surrounding forest district. During the winter, lookouts would hike down to the ranger station to survive.

The CCC and the forest service constructed the nearby stone and wood lookout tower on Yellow Mountain in 1934. It is a fourteen-foot-square stone tower supporting a rebuilt fourteen-foot-square wood cab and catwalk, according to the Edwards-Pitman study, which researched several lookout towers in western North Carolina. Men from the CCC used local fieldstone with a thick concrete mortar to construct the stone base.

Pratt McClure of Franklin, North Carolina, was the last lookout. McClure, a U.S. Army private during World War II, worked on the tower from 1960 until it was decommissioned in 1969, according to genealogy records. Throughout much of Yellow Mountain's manned lifespan, lookouts, including McClure, had to hike up the mountain to reach the tower. A road was finally built up to the tower in the late 1960s. In the 1980s, the forest service, with help from the local community, restored Yellow Mountain's deteriorated cab.

In about 1937, the CCC began work on the Paris Mountain Fire Tower near Greenville, South Carolina. It was one of several projects the CCC undertook in the state, including refurbishing towers at Sassafras Mountain and Corbin Mountain, according to state forestry commission records. The land at Paris Mountain, originally owned by the City of Greenville, was acquired in 1935 by South Carolina to be used as a state park. Paris Mountain was one of sixteen parks in the state to be built by the CCC. The federal agency, which established a camp on site, constructed several structures, including the tower, at Paris Mountain.

In addition to building the original eighty-foot-tall steel tower at Paris Mountain, the CCC also constructed a road to reach the lookout. The tower's location, however, became problematic. It was not built on top of Paris Mountain, which is more than two thousand feet in elevation. From the top of the original tower, the lookout could not get a 360-degree view because the mountain blocked part the view. According to the state park service, Paris Mountain is a monadnock, an isolated hill that rises above a plain.

By 1943, the state had decided to relocate the tower to the top of Paris Mountain, thanks to a generous donation by Mrs. Harry Stephenson, according to an annual state report. On the monadnock's peak, officials also constructed living quarters, including a house and a barn, for the watchman. Ruins from the original tower location, including a former four-room house, can be found in the state park.

Due in part to President Roosevelt's Tree Army, it appears that residents became more aware of the dangers of fires in the 1930s and 1940s. In addition to constructing towers, the CCC sought help from local residents in the Blue Ridge to preserve the environment. "One of the more important roles the CCC camp played in the local community was educating the public about how they can help take care of the environment along with them," Kayla Carter said in her article about Camp Cordell Hull in Unicoi, Tennessee.

As a result of the CCC's public outreach, there were fewer fires ignited in the forests illegally, according to multiple sources. But during the same time, the development of recreation lands throughout the region, including the Great Smoky Mountains National Park, led to additional fires being set accidentally. There was a significant increase of campers, and subsequently smokers, campfires, cooking fires and fires ignited by passing vehicles.

Forest Chief Ferdinand Silcox, who worked with the president to organize and manage the CCC, developed a new firefighting policy in 1935. With CCC men building and manning fire towers, Silcox created the "10:00 a.m. policy," which required all wildfires to be extinguished by 10:00 a.m. the day after they are first spotted. If they were not extinguished on time, the forester could declare an emergency and use a "blank check" to fight the fires. Because the nation still believed that fires had to be fully suppressed, Silcox's policy remained in use for about four decades. By the 1970s, the forest service had begun to reconsider the importance of fire and forest management.

CHAPTER 2

THE FIRE TOWER

The iconic fire lookout tower may possibly be the most revered structure found in America's forests. They are wood, steel and stone structures rising above the nation's highest peaks. In their heyday, there were dozens of fire lookouts in the Blue Ridge Mountains, from the highlands of north Georgia to the mountains of southwest Virginia. And while a vast majority of towers are no longer used for fire detection, the lookouts of the Blue Ridge remain symbols of conservation.

Like their lighthouse counterparts along the nation's coasts, fire lookouts have become beacons in America's mountainous landscapes. Navigators, like ships at sea, use the lookouts to find their way across the mountains.

During the early and mid-twentieth century, dozens of fire towers were constructed in the Blue Ridge Mountains. Based on Daniel Adams's earlier findings in the highlands of Arkansas, lookouts were placed in the Blue Ridge about ten to fifteen miles from one another. The goal of the nation's foresters, who simply wanted to spot fires rather than become architects and builders, was to have as few towers as possible to cover as much area. Towers were designed to cause little change to the landscape of the nation's forests. Where land features blocked views of towers, another lookout was constructed.

According to the U.S. Forest Service, the first known forest fire tower in the United States was built in 1902 on private lands in Idaho. The wooden tower replaced a woman lookout who mounted a horse to watch for fires. The first tower in the eastern United States was built in the Adirondack Mountains of

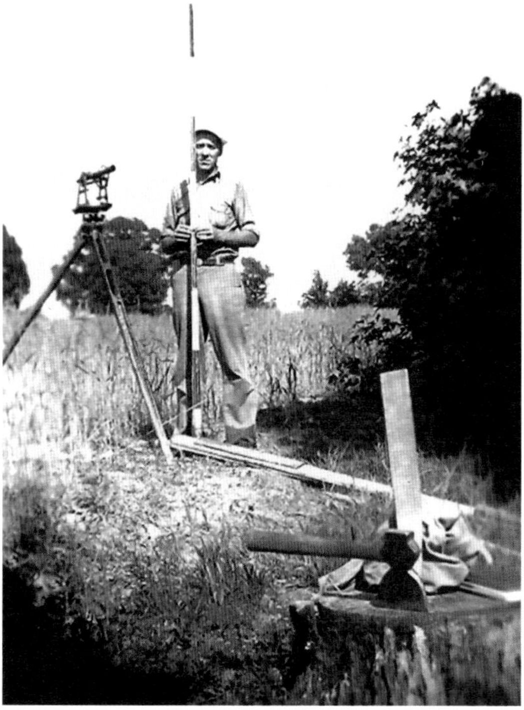

Early surveyors used a variety of tools to locate perfect sites for fire lookout towers in the Blue Ridge Mountains. *Betty Hodge.*

New York in 1909. The U.S. Forest Service noted that it built its first fire tower in 1915 in Oregon.

Like elsewhere in the country, in the Blue Ridge, lookout points were established on the highest peaks in the region. Individuals from the U.S. Forest Service, U.S. Geological Survey, Bureau of Land Management and other agencies surveyed the land and chose locations that were not obscured by other landforms and trees. Although the goal was to find non-obscured sites, there were a few towers at which 360-degree views were not possible.

As the saying goes in the real estate business, foresters selecting tower sites think "location, location, location." In northeast Tennessee, one of the earliest fire towers was built on top of 6,285-foot-high Roan Mountain, one of the highest peaks in the region. Foresters then selected other area high peaks for tower sites, like Holston Mountain, Buffalo Mountain and Stone Mountain, all of which are above 3,500 feet.

Some other high-peak lookout towers were built on Brasstown Bald, Georgia's highest point at 4,784 feet; Mount Mitchell, North Carolina's highest point at 6,684 feet; Mount Rogers, Virginia's highest point at 5,729 feet; Sassafras Mountain, South Carolina's highest point at 3,554 feet; and Clingmans Dome, Tennessee's highest point at 6,643 feet. Lookout towers were also built on other high peaks, such as Clinch Mountain, Mount Sterling, Rabun Bald, Bearwallow Mountain, Mount Pisgah, Mount Cammerer, Pine Mountain and High Knob.

The higher peaks required shorter towers. Some of the lookouts on the highest peaks were simply one-room cabins and stone structures, such as the short stone lookout on Mount Cammerer in the Great Smoky Mountains National Park and a short two-story stone and wood lookout on Rabun Bald in north Georgia. There have been several towers on Virginia's High Knob, but none was very tall. The first was a one-room cabin. Two- and three-story lookouts were later built on High Knob. But even on some of the higher peaks, there were some very tall towers. The tower on Holston Mountain was one hundred feet tall, much higher than what was actually necessary. Holston's lookout was originally built in Mississippi, where the elevation was much lower. It was moved to Tennessee after the original wood Holston tower burned. A tower of similar height was built on Roan Mountain, but foresters learned that at the peak's high elevation, a tall tower was inadequate. A tower on nearby White Rock was used more than the short-lived tower on Roan Mountain. Tall steel towers were also constructed on Mount Sterling, Sassafras Mountain and Bearwallow Mountain.

Early towers, built of native wood, were usually constructed to fit the landscape. There weren't any standard blueprints for wood towers. Speculators would find the perfect location to build a tower, and construction crews would build the tower depending on the surrounding landscaping. Crews usually hiked in, set up camp and used local materials to build the towers.

Joanna Bald Lookout in Graham County, North Carolina, which is now a steel tower, was originally built as a wood structure in the 1930s. The original tower looked more like a small church chapel than a fire lookout, according to old photographs of the site. It featured a white cupola, shuttered windows and a small front porch. By the 1950s, a steel tower had replaced the wood structure, according to a forest service engineering study.

A wood cab once stood on top of 4,102-foot Comers Rock along the Wythe and Grayson county line in Virginia. The nineteen- by fourteen-foot cab was constructed on a stone foundation in about 1930 in what is now the Mount Rogers National Recreation Area. With windows wrapping around the building, the lookout could get a 360-degree view from inside the cab, or he could just stand outside on the rock outcropping and take in the incredible view.

Shortly after the cab's construction, the CCC established a firebreak at Comers Rock. A firebreak is a gap in vegetation used to slow or stop fire. The firebreak eventually became Jefferson National Forest Road 57, which accesses the lookout site. The lookout, which was sold in the 1950s, has been removed from Comers Rock.

When the government began replacing the wood structures with steel towers, the forest service developed detailed plans. Depending on the

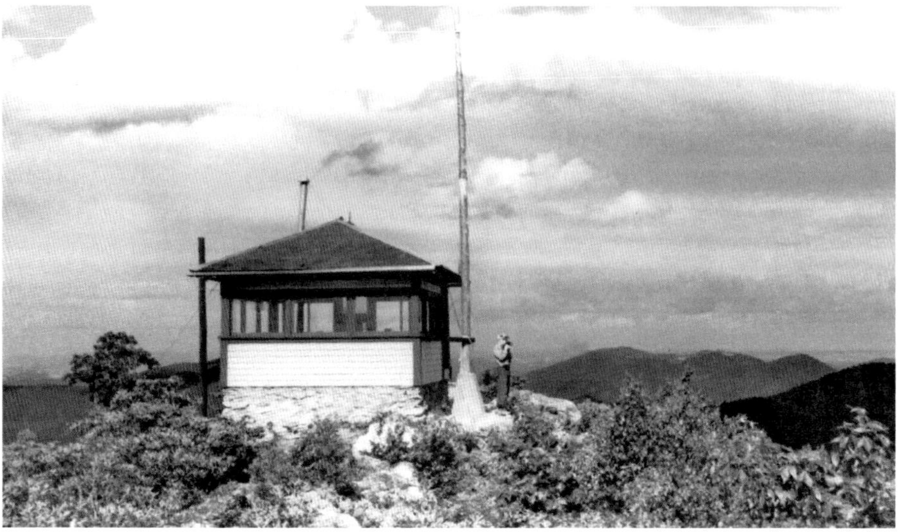

Comers Rock was a lookout cab high above the mountains in Wythe County, Virginia. It has long been dismantled. *National Archives.*

landscape, different blueprints were used. Some towers featured live-in cabs on top of the towers, and at others, the lookouts spent the day on the tower and slept in a cabin on the ground at night.

During the mid-1920s, Forest Chief William Greeley published two volumes containing forest service–approved specifications. The specifications included plans for the construction of lookout towers and lookout cabins. Greeley's report, according to the Edwards-Pitman study, called for two galvanized-steel tower plans. There were very few differences between the two tower designs, with the location of the ladder being the main difference. One called for the ladder to be located on the exterior of the tower, and the second called for the ladder to be located on the interior.

"In twelve years experience with outside ladders on steel towers no lives have been lost and, as a matter of fact, no minor accidents resulting from the use of that type of ladder are of record," Greeley wrote in his 1924 report, according to the Edwards-Pitman study.

Lookout cabs in the Blue Ridge follow a basic pattern, as the Edwards-Pitman study has noted. The fourteen- by fourteen-foot cab—featuring windows on all sides, minimum obstruction of the view by the corner posts and window frames and a catwalk with guardrail—included a four-panel hip roof. The study noted that each tower also featured lightning protection and grounding, as the towers had to withstand storms passing through the mountains.

Two companies from the midwestern United States—the Aermotor Company and International Derrick and Equipment Company—manufactured many of the towers still standing in the Appalachian Mountains. Manufactured of sturdy Midwest steel, the two companies built the lookouts in coordination with the U.S. Forest Service, which designed the structures.

The Aermotor Company of Chicago, Illinois, manufactured a majority of the steel towers in the Appalachian Mountains and based those towers on Greeley's plans. The company also manufactured many of the towers in the Adirondacks and Catskill Mountains of the Northeast, as well as towers throughout the South and Midwest. Not many, if any, Aermotor towers are known to have been constructed in the western United States.

Aermotor is likely better known for its windmills, which can be found around the world, but it also manufactured fire lookouts based off the windmills. The company started building the all-metal towers for the forest service starting in the 1920s. Its towers were installed in the Blue Ridge during the 1930s and 1940s. And because they are based on the windmill structures, the fire towers feature small cabs, usually about seven by seven feet.

The Bearwallow Mountain Lookout in Henderson County, North Carolina, is an excellent example of an Aermotor tower. The unique lookout tower was erected and manned by the North Carolina Forest Service in 1934. With metal painted yellow, the tower stands forty-seven feet above the mountain and features a seven- by seven-foot cab. From the cab, the lookout could see all of Henderson, Buncombe and Rutherford Counties. Located on a barren bald grazed by cattle, a forest service cabin is next to the tower. Bearwallow was manned until 1990, according to the Forest Fire Lookout Association. The Aermotor tower remains in good condition in the twenty-first century. There are a few communication antennas now on the mountain with the tower.

The International Derrick and Equipment Company—with locations in Columbus, Ohio, and Torrance, California—also manufactured fire towers that can be found throughout the Blue Ridge. The company manufactured two tower designs—one in 1933 was based off the Aermotor towers, and in 1937, the company manufactured a tower designed by the forest service. International Derrick towers, which were of sturdy build, were constructed in the Great Smoky Mountains National Park at Shuckstack Mountain in 1934 by the Public Works Administration and at Mount Sterling in 1935 by the CCC.

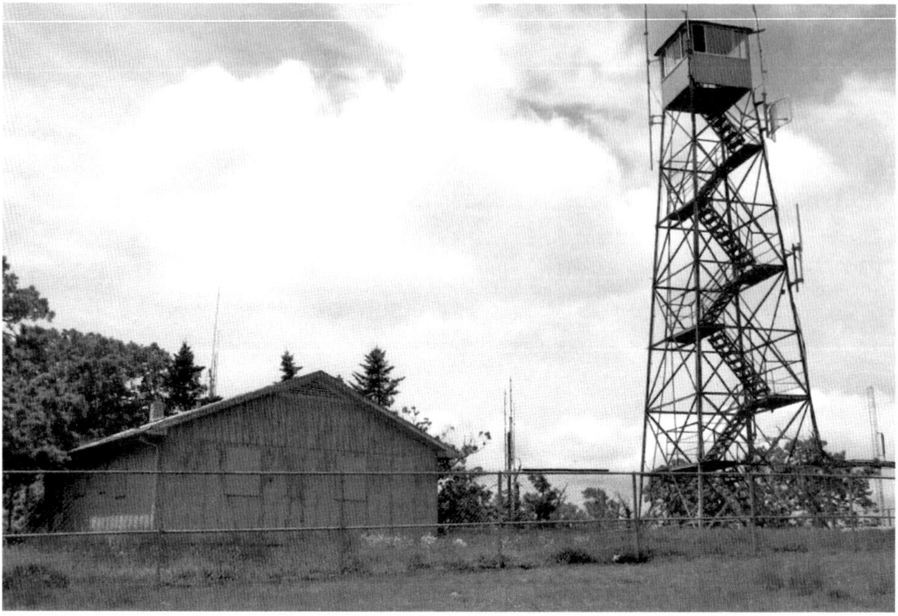

An Aermotor steel tower stands on top of Bearwallow Mountain in North Carolina. Cattle regularly graze on the bald. *Photo by Robert Sorrell.*

Shuckstack, a sixty-foot-tall tower in Swain County, North Carolina, has recently been the subject of the Save Shuckstack campaign. Although of sturdy build, the tower's bolts are rusting, the floor planks are rotting, the roof is littered with holes and the guardrails are missing. "The other three remaining towers [including Shuckstack] are cultural resources, but do not have any special designations bringing them to the top of our restoration priorities at this time," said Dana Soehn of the Great Smoky Mountains National Park, which does not recommend climbing Shuckstack.

The International Derrick Company also built the fire lookout on Mount Sterling, another tower in the Smokies. Located in Haywood County, North Carolina, Mount Sterling's tower, with a 7- by 7-foot cab, was assembled in 1935 by the CCC. It is a 60-foot-tall tower at 5,835 feet above sea level, the highest steel fire tower in the Blue Ridge Mountains. Like many remote lookouts in the Great Smoky Mountains National Park, the tower on Mount Sterling is not easy to get to. It is located about three miles from the nearest service road. Its cabin, which no longer stands, was built one year before the tower, according to national park records. International Derrick also built the 31-foot-high steel tower on

Joanna Bald, which replaced an original wood lookout. Joanna Bald's tower features a 14- by 14-foot live-in cab.

Many Aermotor and International Derrick towers feature the Carnegie brand stamp. The Carnegie Steel Company, which existed from the late nineteenth to the mid-twentieth centuries, manufactured the steel used to build the towers. Andrew Carnegie's company, based in Pittsburgh, Pennsylvania, built a variety of steel products, including steel beams. Carnegie steel was used to manufacture the tower at Mount Sterling. The Bethlehem Steel Corporation of Pennsylvania also manufactured steel that

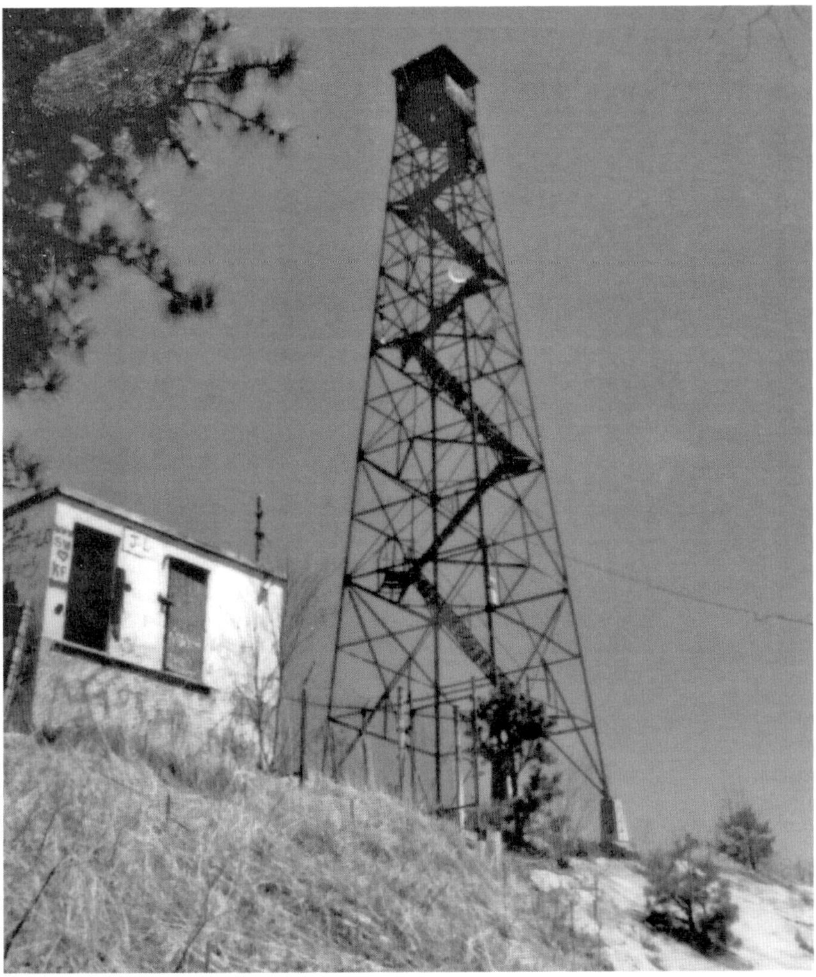

A steel Aermotor fire tower has stood on Clinch Mountain in Mendota, Virginia, since the 1930s. *Photo by Robert Sorrell.*

was used to build towers. The three-stage tower at Rich Mountain is made of Bethlehem steel. Aermotor and International Derrick stamps can also be found on towers.

Because Aermotor and International Derrick towers featured small cabs, lookouts spent the night and cooked in a wooden cabin on the ground. Shuckstack and Mount Sterling had small one-room log structures.

Another steel tower, Mendota Fire Tower, is located in Washington County, Virginia. The nearly one-hundred-foot-tall tower features a small lookout cab. Unfortunately, Mendota's fire tower and an adjacent block structure have been heavily vandalized over the years. The bottom flight of steps of the tower has been removed to keep people off the lookout.

Several steel towers have risen over the Blue Ridge Mountains of South Carolina, including lookouts at Paris Mountain and Sassafras Mountain, the state's highest point. The state's forestry commission constructed a steel tower on Sassafras Mountain, which actually straddles the state line with North Carolina, in the 1940s. The tower stood on the mountain's peak for a few decades but was eventually dismantled. It was one of the

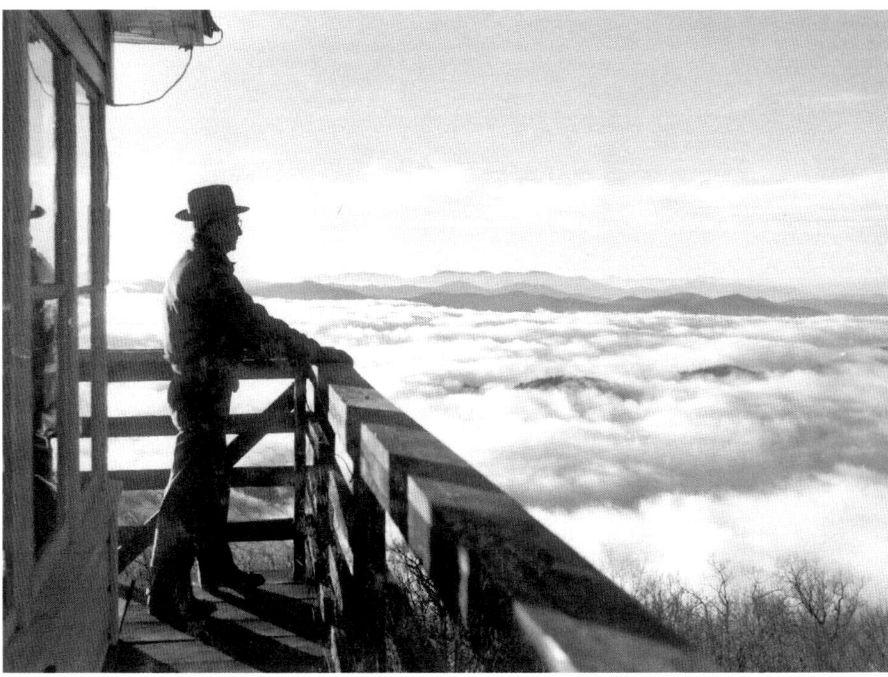

A man watches for smoke over the terrain below the Black Mountain Fire Tower in north Georgia. *National Archives.*

state's shortest towers at fifty-five feet tall, about half the size of towers in South Carolina's lowlands. But from the steel tower, the lookout could see forests in four states.

North Georgia, which is better known for its stone lookout towers, has had a few steel towers. The thirty-three-foot-high tower at Black Mountain was constructed in 1949. The steel tower, located near Ranger Arthur Woody's home in Suches, is relatively close to Brasstown Bald. But unlike Brasstown

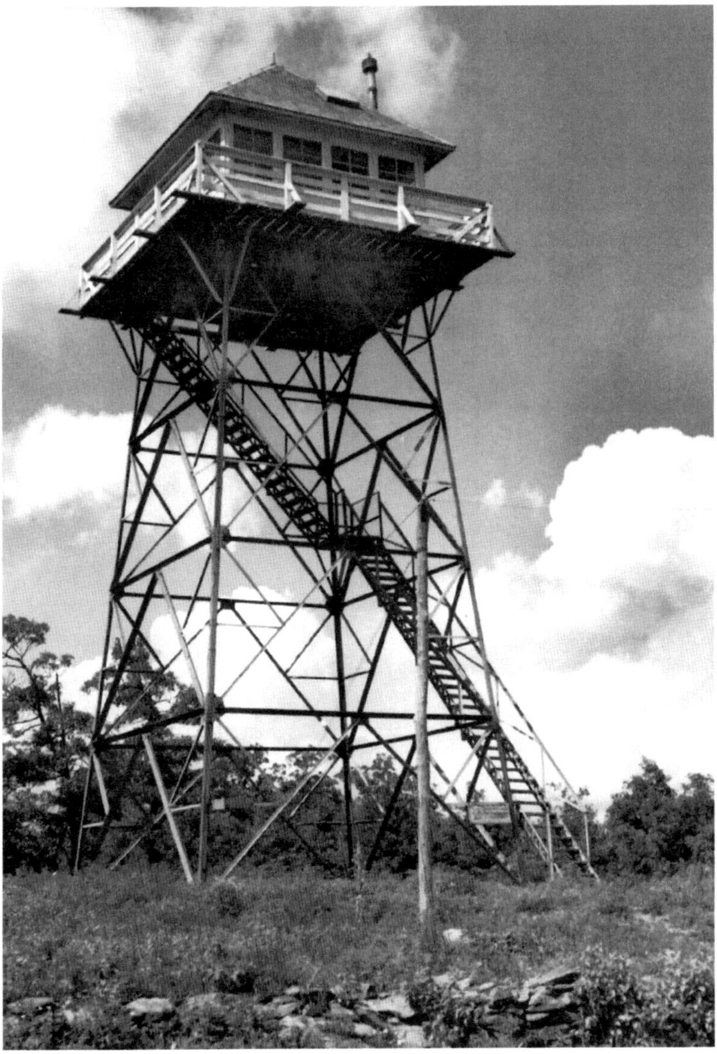

The Grassy Mountain Fire Tower, a steel lookout structure, is located in the Chattahoochee National Forest of north Georgia. *National Archives.*

Bald, Black Mountain's tower was a steel structure, with a thirteen-foot-square live-in wooden cab with a catwalk. The structure is located on the summit's black rock surface.

According to the Forest Fire Lookout Association, the Chattahoochee National Forest in north Georgia is home to similarly named towers at 3,315-foot Glassy Mountain and 3,511-foot Grassy Mountain. The tower at Grassy Mountain, in Murray County, was built in 1938. The 50-foot metal tower features a 14- by 14-foot metal cab with a catwalk. The 55-foot Glassy Mountain tower was built in 1941 in Rabun County. It features a slightly smaller 13- by 13-foot metal cab with catwalk. While Grassy Mountain's tower is surrounded by a chain-link fence, the forest service built a ground residence and garage around Glassy Mountain in the 1960s.

Mount Cammerer in the Great Smoky Mountains National Park, Wayah Bald in southwestern North Carolina and High Knob Tower in southwest Virginia were built of stone and rock, as well as wood and steel. These towers featured a variety of designs, similar to some of the stone lookouts constructed out west. The CCC built a handful of stone lookouts during the 1930s in the Blue Ridge, including at Mount Cammerer and Wayah Bald.

Other stone lookouts were built at Rabun Bald, Brasstown Bald, Chenocetah Mountain, and Fort Mountain in Georgia, as well as Mount Mitchell in North Carolina, which has been home to at least five lookout structures since the Civil War.

A monument and makeshift lookout were established on Mount Mitchell in 1888 after the death of namesake Elisha Mitchell. By 1915, another makeshift lookout, with a ladder, three poles and a platform, had been was erected. By 1916, when Mount Mitchell State Park was established, the state had erected a fifteen-foot-high covered wooden lookout on the summit.

A 1925–26 biennial report from the North Carolina Department of Conservation and Development noted that during the fall of 1925, a steel tower was erected on the summit of Mount Mitchell, which at 6,684 feet above sea level is the highest point east of the Mississippi River. Before it was completed, however, a heavy sleet storm caused the steel tower to collapse. The steel tower was to be used for fire detection. "Through the generosity of Col. C.J. Harris, of Dillsboro, a rough stone tower has been promised to replace the steel structure, and steps have already been taken to complete the building during the summer of 1926," the report said.

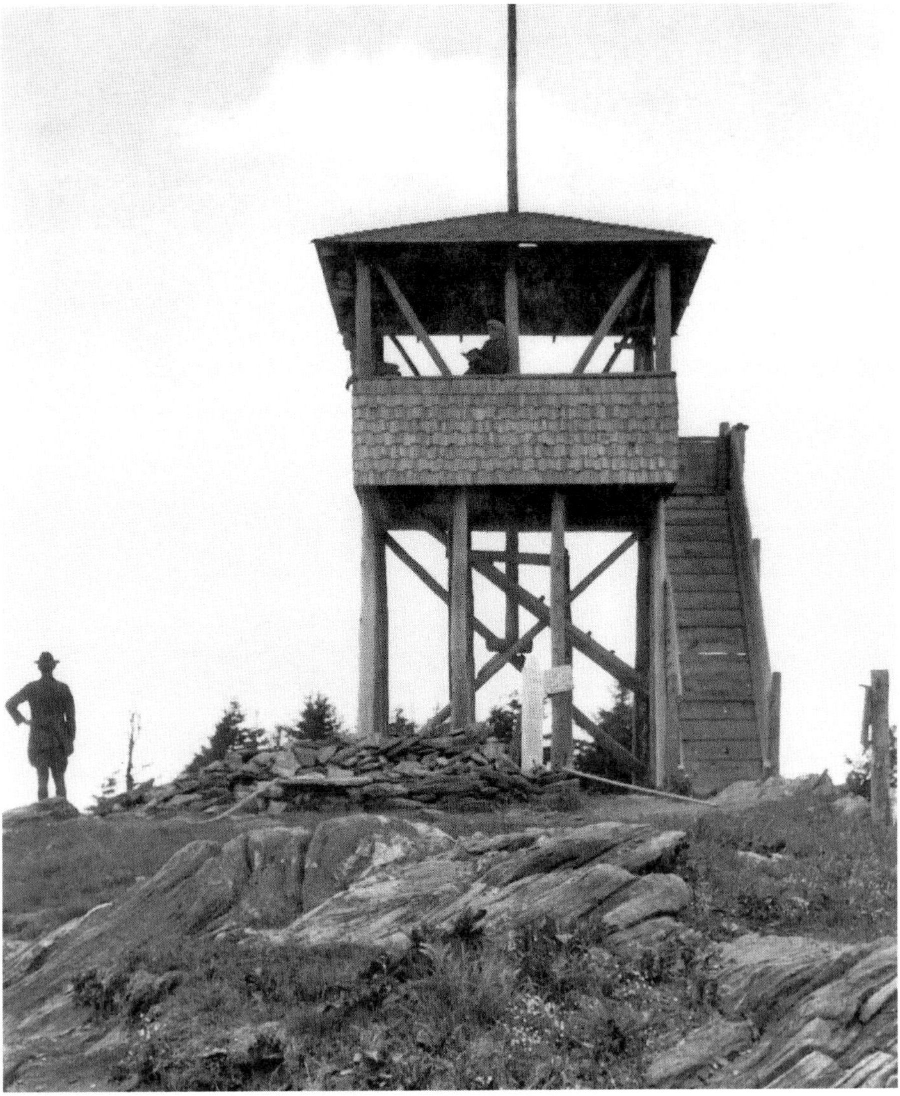

Several lookout towers have been constructed on Mount Mitchell in western North Carolina, including this wooden lookout tower. *National Archives.*

Completed in 1926, the stone tower features a medieval motif and stood on Mount Mitchell for nearly four decades, until 1960, when a nearly thirty-five-foot-high stone tower with an observation deck was built. The second stone tower was demolished in 2008 and replaced with a lower observation lookout.

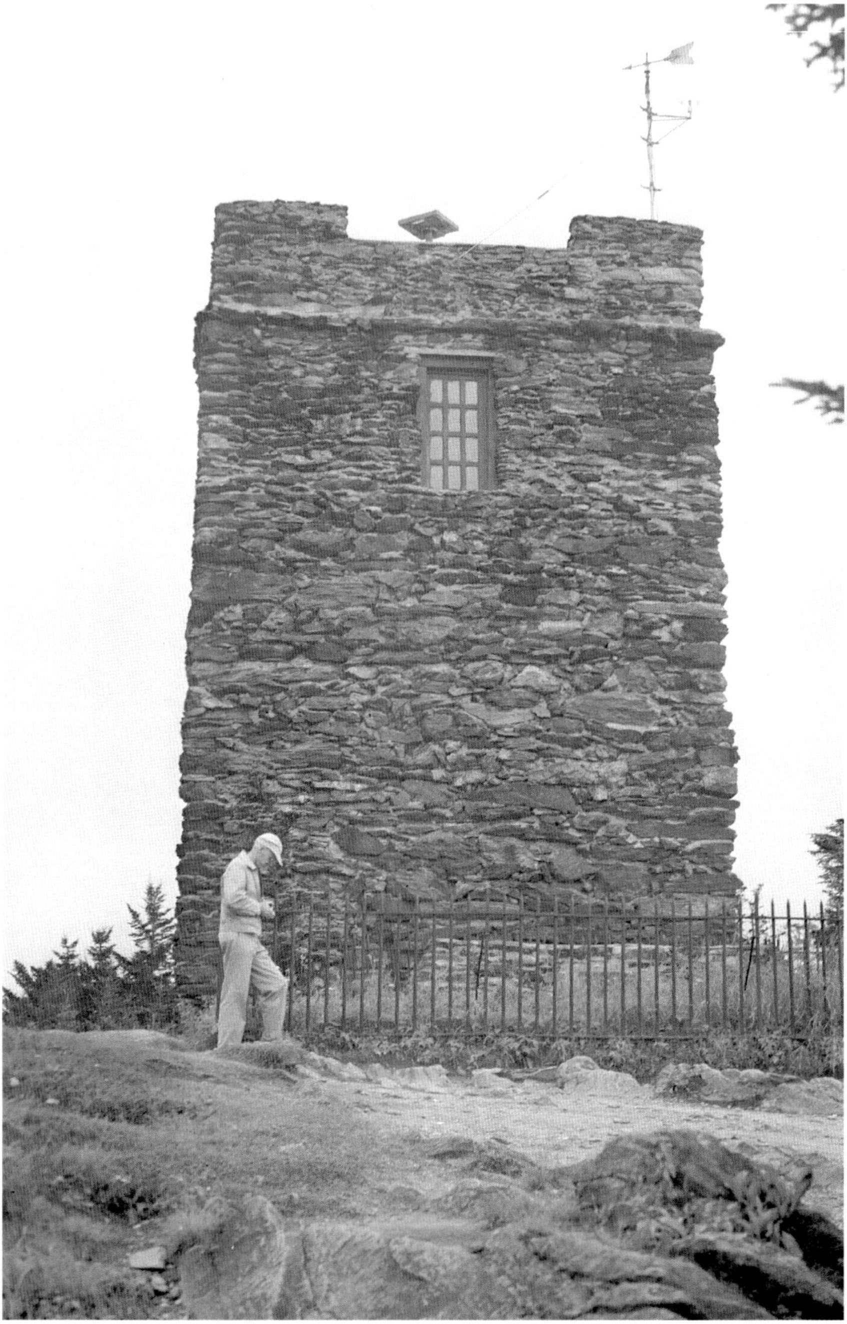

Several fire lookout towers have been built on Mount Mitchell, the highest point east of the Mississippi River, including this stone structure. *Albert "Dutch" Roth Collection, University of Tennessee.*

"To stand on Mount Mitchell's summit and enjoy the breathtaking scenery from the highest peak in eastern America is one of the most satisfying traditions of our state parks system, and this new facility allows the park to continue to offer that experience," said Lewis Ledford, North Carolina state parks director, in 2009. "Observation structures on the mountain have been part of its history, and it was important to develop a design that is in harmony with the area's beauty, fits comfortably with the natural contours of the mountain and is accessible to all visitors."

Fort Mountain State Park in north Georgia is home to the CCC-built Fort Mountain stone fire lookout. The thirty-eight-foot stone fire tower was built on the peak of Fort Mountain around 1935. At the time, the state was in the process of developing a state park on the mountain. According to the state, Fort Mountain's tower is similar in design to the original stone tower at Brasstown Bald. Fort Mountain State Park, which is listed on the National Register of Historic Places (NRHP), also features a unique ancient rock wall that has captivated archaeologists since its discovery in the early twentieth century.

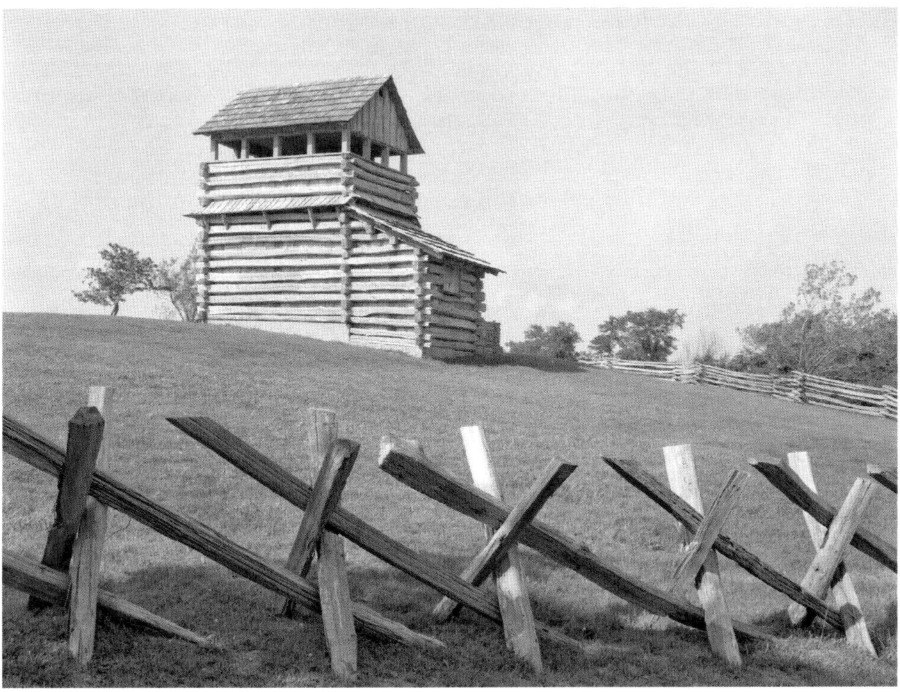

The Groundhog Mountain Fire Tower, a wood log lookout building, is located along the Blue Ridge Parkway in Carroll County, Virginia. *National Archives.*

The fire tower on Groundhog Mountain in Carroll County, Virginia, is a one-of-a-kind lookout. The tower is located along the Blue Ridge Parkway at milepost 188.8. Constructed in the 1930s by the Virginia state forest service, the tower was built of logs to look similar to an old tobacco barn. The 3,025-foot-high Groundhog Mountain property is surrounded by log fences, which were built in 1939.

While a tower structure is important, a lookout cannot do his or her job without the proper tools of the trade. The fire lookout cab, an enclosed room on top of the tower usually at least seven feet wide, is used to house a variety of tools and equipment. Typical items inside the cab include an Osborne fire finder, the lookout's most important tool; monoculars; binoculars; a two-way radio; a telephone; recordkeeping journals; a small table; and a stool. Cabs used as living quarters also included a bed or cot and a stove for cooking.

In the Blue Ridge, where the forest service issued a set of plans for the tower's structure, it also laid out requirements for placement of items in the cab's interior. The typical Blue Ridge tower, with a fourteen- by fourteen-foot cab, had a fire finder stand in the middle of the room. From this point, the lookout could see through every exterior window.

The Edwards-Pitman study noted that the cab's layout was designed for the comfort and convenience of the lookout staff. The typical cab featured built-in cabinets along the interior walls. A bed or bunkbed, a stove, a water stand, a cooler cabinet and a table usually lined the walls of the cab.

The Osborne fire finder, which was mounted on a stand in the middle of the cab, was developed in 1915 by William "W.B." Osborne. He was a forest service employee from Portland, Oregon, and designed the fire finder so wildfires could be precisely pinpointed. The instrument is a type of alidade used in order to find a directional bearing, or azimuth, for smoke in order to alert fire crews to a wildfire.

Many fire finders were manufactured from 1920 through 1935, and the manufacturer, Leupold & Stevens Inc., stopped production of replacement parts after 1975. Leupold & Stevens has also manufactured rifle scopes, binoculars and other sighting equipment. Recently, the government has worked with Palmquist Tooling Inc., a California company, to manufacture the fire finder. Palmquist is the sole manufacturer of fire finder equipment today.

The fire finder is composed of a topographic map of the area around the tower, oriented and centered on a horizontal table, or stand, with a circular rim graduated in degrees. Two sighting apertures are mounted above the map on opposite sides of the ring and slide around the arc.

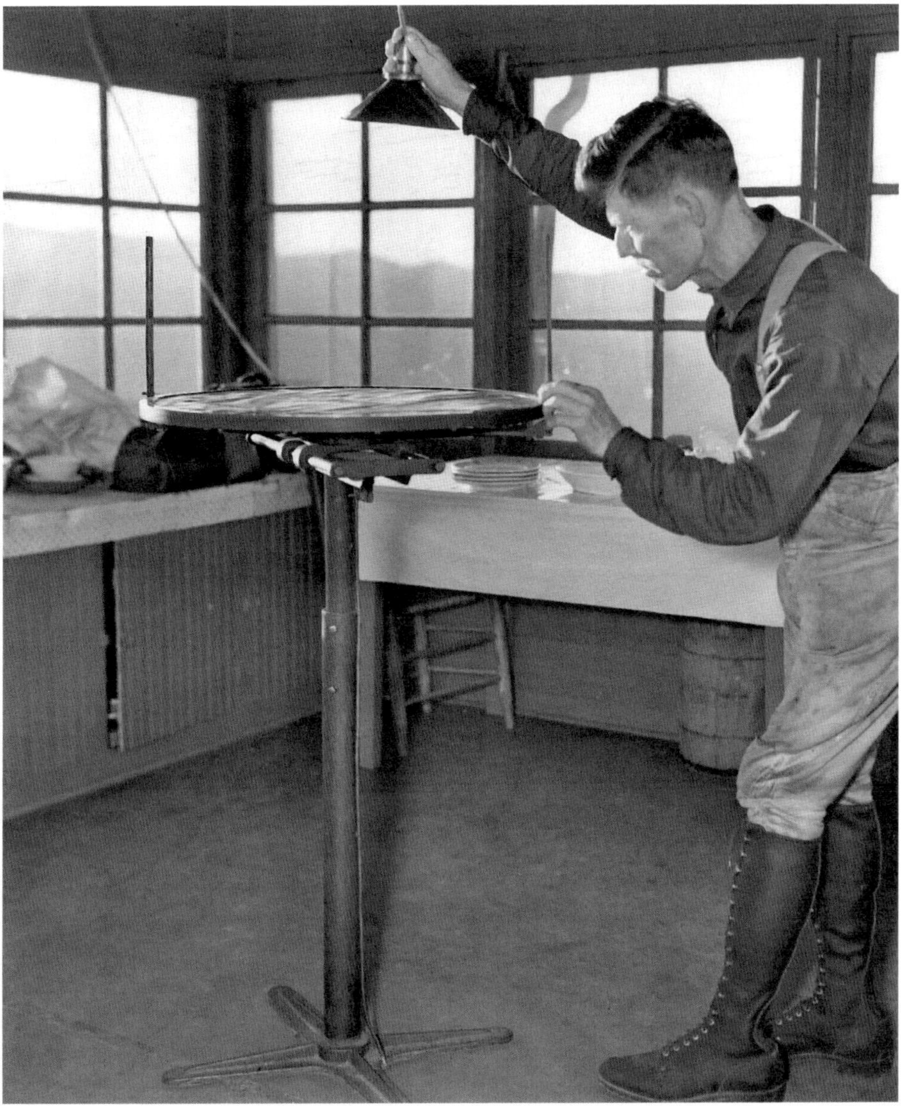

V.F. Colin sights smoke through an Osborne fire finder in a state-managed lookout tower near Tellico Plains, Tennessee. *Tennessee State Library and Archives.*

According to the forest service, the device is used by moving the sights until the lookout can see through the nearer sighting hole and the cross hairs of the further sight align with the fire. The fire lookout notes the degrees on the graduated ring beneath the sight. It typically takes at least two fire lookouts to precisely pinpoint a fire.

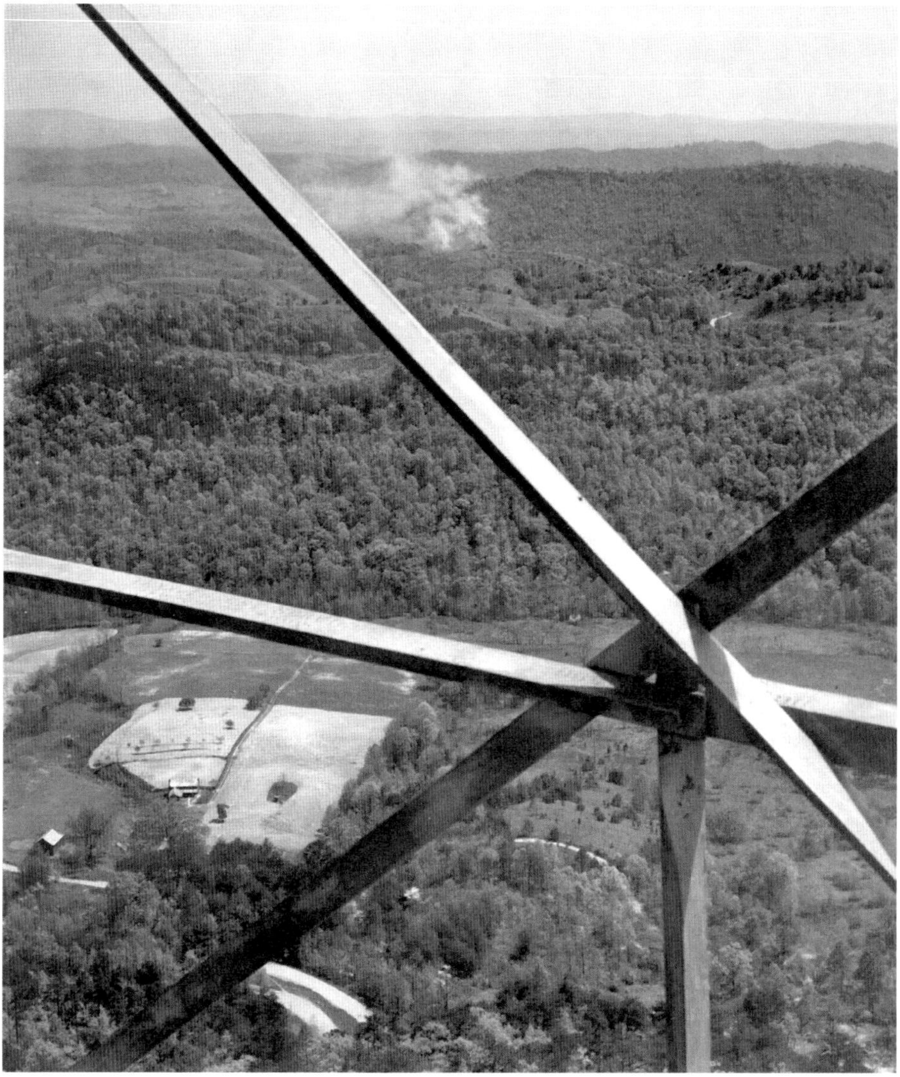

Smoke billows from a forest fire on Clinch Mountain, as seen from the Little War Gap Fire Tower in Hawkins County, Tennessee. *Tennessee State Library and Archives.*

Osborne also designed a camera, known as the Osborne photo-recording transit, which was used to take panoramic photographs. According to the National Park Service, the Osborne camera was custom made and combined the features of an engineering surveyor's transit with photographic capabilities. The seventy-five-pound camera was carried on the lookout's back as he or she climbed the tower and was used on the catwalks (decking area around the cab) and roofs.

After sighting smoke or flames from the tower, the lookout communicates with fire dispatchers at a central facility, usually at a nearby town, a ranger station or at another tower. Dispatchers would work with multiple fire lookouts to determine the exact location of a fire. Firefighting personnel would then be dispatched and directed to the blaze.

As fire tower designs have changed, so have means of communicating with dispatchers. The earliest lookouts sent messages to dispatchers by carrier pigeons. The forest service used this old-fashioned communication method until the mid-1920s.

In 1919, the Ellensburg, Colorado *Evening Record* newspaper reported that the forest service was preparing to use carrier pigeons. The pigeons, from the navy department, could travel distances of up to six hundred miles, the article noted, but the lookouts were less than fifty miles from firefighting areas. "The value of the birds in the mountainous and lake regions where in many places travel is most difficult, and where there are but few telephone and telegraph lines, will be especially great, it is believed."

Lookouts also used heliographs to signal dispatchers in the 1910s. The system used one or two tripods, two mirrors and a shutter to flash a message in code. The message could be seen as far as fifty miles away if there was enough sunlight. But there were a few obvious problems with the heliograph system. First, there had to be someone to watch for the code. The codes also had to be memorized and easily understood. The weather also had to cooperate.

In 1920, William Norman Millar wrote about the use of heliographs by the forest service in *Methods of Communication Adapted to Forest Protection*. Next to the telephone, Millar said, the heliograph was, at the time, the best method of communication for forest protection. He noted in his report that use of the heliograph was limited in the eastern United States.

The most common means of communication, the telephone, was first established by the forest service in 1906 on a reserve in Wyoming. From then on, many rural communities across the country received their first phones because of the forest service's fire detection network. The massive fires of 1910 in the northwestern United States emphasized the need for better communications in rural America, particularly within the forests. As a result of the fires, the forest service developed a telecommunications network around the country and looked into other methods of communication, such as radio.

Because the forest service's phones were the first phones in many locations, the agency entered into private phone contracts. Residents were allowed to use forest service phone lines in exchange for working as fire patrol personnel. Commercial telephone systems developed later.

Both Fred Stout and Murphy Potter of Carter County, Tennessee, worked on the former White Rock Fire Tower, according to Emma Ruth Shoemaker. The family's home had the first telephone on Buck Mountain, a rural mountain community near the state line. Shoemaker said that having a phone at home allowed the men to keep in touch with fire officials.

George Osborne, who now hikes regularly to the Pinnacle Mountain Tower in Unicoi County, Tennessee, remembered the historic Buckner Ridge Fire Lookout in Virginia. Osborne grew up in the Fort Blackmore community of Scott County, Virginia, and recalled visiting the old tower, which has been removed.

Osborne's uncle lived with his family when he worked on Buckner Ridge. The Osborne house was the first home in Fort Blackmore to have phone service because of the forest service. His uncle allowed him to visit the tower, which was coordinated with lookouts at Mendota, Hayters Knob and High Knob. Most of the forest service phone-related equipment was based on American Telephone and Telegraph Company equipment or was purchased from the company.

In addition to building fire towers, the CCC also constructed much of the forest service's phone network. The phone lines and poles were constructed along the jeep roads and trails, so crews could easily make repairs. Every spring, the nation's forest districts conducted maintenance checks on the phone lines. The lines were designed to work through snow, strong winds and fallen trees. Most lookouts used wooden wall phones, but portable phones and other telephones were also utilized. The forest service also utilized emergency telephone lines.

In South Carolina, which started using telephone lines in about 1930, tower lookouts would even climb the telephone poles, tap into the lines and make reports back from the field. The state, which had more than two thousand miles of telephone lines, disposed of the system around 1970. Many of the lines were eventually purchased by local phone companies and cooperatives.

One interesting story about phones and the South Carolina Forestry Commission comes from the Upstate. When the lines were being built, a woman in Greenville County agreed to provide a line right-of-way across her property in exchange for the forestry commission's promise that it would provide her service as long as she lived. She outlived the forestry commission's use of the telephone network, but the commission honored the agreement until she died.

In 1941, when the CCC program ended, the forest service began using radios for backcountry communications. Although telephone lines and phones were still used throughout the Blue Ridge, radios became the primary means of communication. Radio usage also saved the forest service money because men were not needed to maintain the telephone system. Radios could also be used with aircraft and crews working on the fire line. Portable radios were used in the towers, in the air and on the ground.

The forest service's radio communications network outlived the fire lookout system. When the forest service decommissioned the fire detection system, many towers were then used to house radio equipment and antennas. Many towers today are still used for radio communications. In Virginia, state police have used former fire towers to house communications equipment.

Access to the tower is another aspect of fire lookouts. During the 1930s, the CCC constructed hundreds of thousands of miles of trails and roads, many of which provided access to the fire towers. Most roads

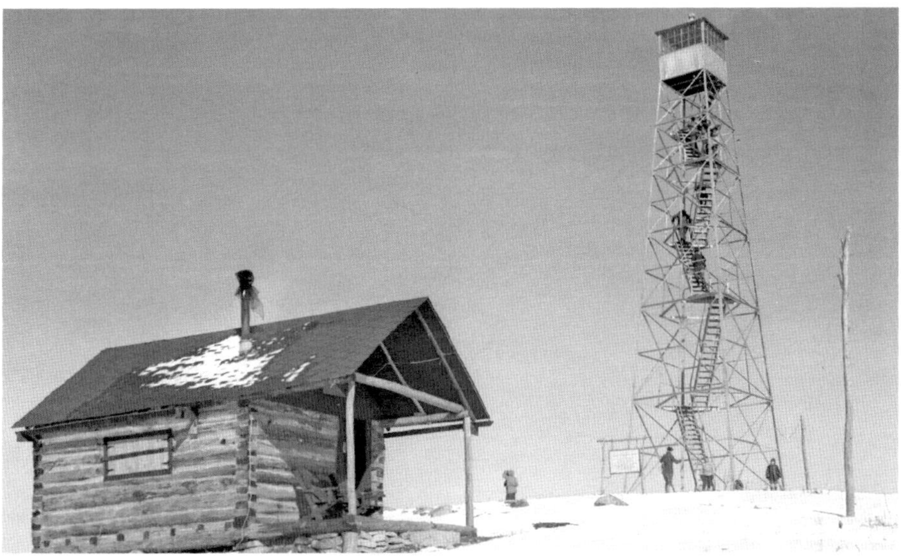

Hall Top Fire Tower and cabin were located in the Cocke County, Tennessee portion of the Great Smoky Mountains National Park. *Albert "Dutch" Roth Collection, University of Tennessee.*

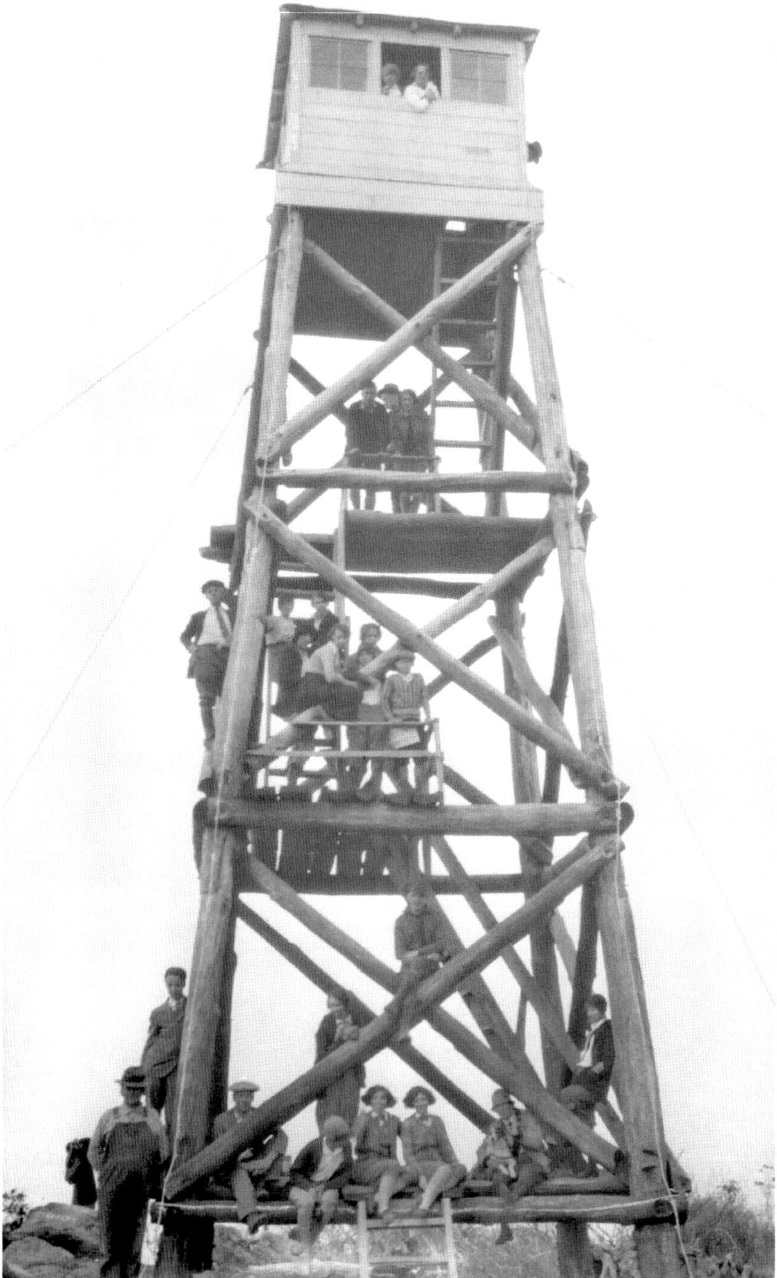

Albert "Dutch" Roth photographed many fire lookouts in the Blue Ridge Mountains, including the tower on Blanket Mountain. *Albert "Dutch" Roth Collection, University of Tennessee.*

were narrow dirt and gravel paths that led up the sides of the mountains to the towers. The CCC would use a variety of tools to build the roads, including horse-powered equipment and heavy machinery. The roads were built so lookouts could travel to the tower by four-wheel-drive vehicles, hence the name of the old jeep roads. They weren't necessarily constructed for family excursions, although some roads allowed visitors to access the mountains. In many locations, such as Yellow Mountain in North Carolina, the CCC also blazed trails and footpaths to the towers. Most mountain roads and trails followed stream beds or former Native American and pioneer paths.

Albert "Dutch" Roth, who lived from 1890 to 1974, is one of the most prolific early photographers of the Great Smoky Mountains National Park. A collection of his photographs is now archived at the Great Smoky Mountains Regional Project at the University of Tennessee in Knoxville. According to the archives, "What began in 1913 as a diversion developed into a serious avocation as Roth perfected his penchant for photography while avidly hiking the unexplored regions near his home."

He'd carry his Kodak camera and a heavy tripod through the mountains along the Tennessee and North Carolina border. He is most recognized, according to the archives, for his photographs of the national park's Greenbrier and Mount LeConte sections. On many occasions, Roth ventured outside the national park to places like Mount Mitchell and Albert Mountain.

Roth's grandson, Charlie Roth, who donated a collection of his photographs to the archives, said that his grandfather loved to hike and explore the mountains. On those hikes, Roth regularly visited the area's numerous fire towers. Between 1924 and 1959, Roth kept a hiking journal, which periodically mentions visiting fire towers. "Mount Mitchell is the highest mountain east of the Mississippi, but you can drive within a mile of the stone tower on top," he wrote in August 1951. "From here you can get a grand view in every direction. You go up in the tower like you would a winding stairway in an old castle tower."

In another passage, Roth wrote about visiting the tower on Mount Sterling: "We got to the top about lunch time and ate beneath the steel tower. After we rested awhile and climbed up to the top of the tower and taken some pictures, we started back. It wasn't bad going back, because it was down hill all the way."

During the 1940s and 1950s, Roth photographed several fire lookout towers throughout the Blue Ridge Mountains. The photographs, now

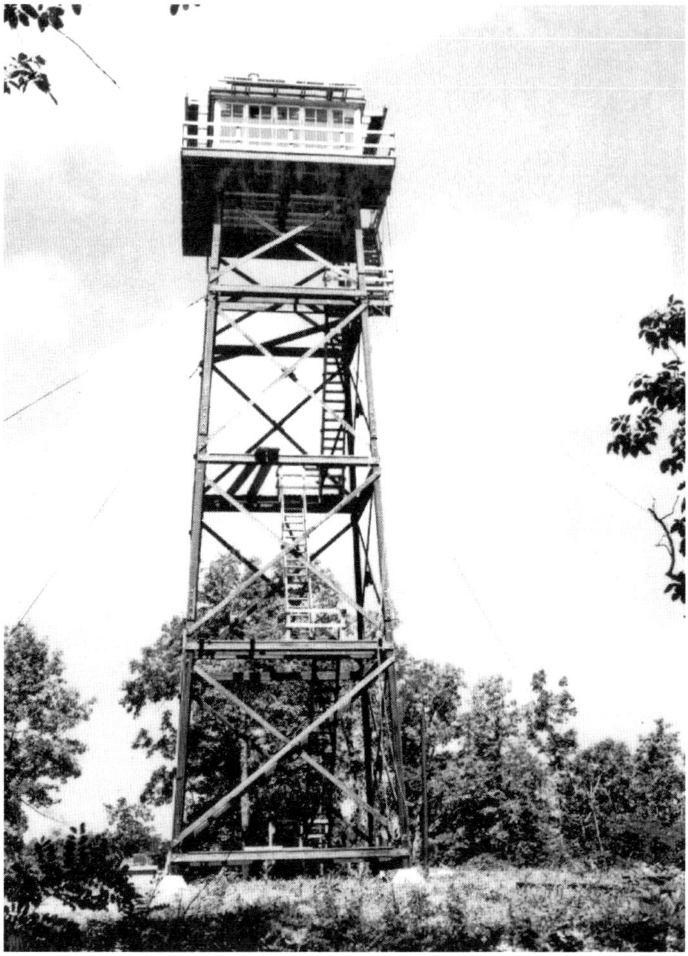

The Chestnut Mountain Fire Tower, pictured in 1941, was one of five towers in the Armuchee Ranger District of the Chattahoochee National Forest in north Georgia. *National Archives.*

featured in the university archives, include early images of Barnett Knob, Albert Mountain, Mount Mitchell, Mount Cammerer, Bluff Mountain, Wayah Bald and Blanket Mountain. Many of the photographs depict towers that are no longer standing, including Blanket Mountain and the original Wesser Bald Tower.

CHAPTER 3
THE LOOKOUT

For more than a century, fire lookouts have enjoyed careers of solitude. Working a job that nearly everyone would want (at least those not afraid of heights), they have manned fire towers across the Blue Ridge Mountains, watching for smoke and flames in the surrounding forests.

The lookout, who occupies the confines of the tower's cab, is a man, a woman, a married couple, a professional forester, an outdoor hobbyist, a teenager, a retiree, a writer or a scientist. Anyone who met the forest service's few requirements could become a lookout.

They spend hours, days and weeks atop the highest peaks in the mountains working on a variety of tasks, most importantly being the eyes of the forest service. With the tools of the trade by their sides, such as binoculars and monoculars, lookouts have been, and still are in many locations, the first point of battle for fire suppression in the nation's forests.

"Contrary to what most people think, the men who man the lookouts lead very busy lives while performing their duties. They are so busy, in fact, that they do not have time to get lonesome," noted a writer in *Warwoman Camp News*, a CCC newsletter in north Georgia.

Typically, fire lookouts in the Blue Ridge worked during two seasons: spring and fall, with fall being considered the driest time of the year. Some lookouts worked one or two fire seasons, and others worked for the forest service for decades. At many towers, the lookouts were called on when needed and in case of emergency.

According to the Department of the Interior's *Lookout Manual,* the job description is clear: "You are the eyes of the fire detection system and the quick suppression of fire depends largely on you reporting a fire as soon as it starts. Your motto should be 'constant vigilance and alertness.' Remember that every day is Fire Day."

Another manual, from the U.S. Forest Service in 1916, noted that lookouts must be precise and quick. According to the manual, report time should not exceed eight minutes: "It may take longer than a three-minute period for the lookoutman to determine whether the smoke is from a fire which must be fought, but within three minutes from its appearance he should have its bearings and start to report, so that the dispatcher and fireman may be getting ready…If a lookoutman seems slow or inattentive he should be checked up with test fires."

The dispatcher is responsible for preparing and moving firemen to fight the blaze. In some circumstances, the dispatcher participated in firefighting. Lookouts were required to be in good health, have good eyesight and possess a clear speaking voice. Many lookouts were hired from within the forest service and were familiar with fire control, road and trail construction and maintenance, timber cruising and surveying.

Good health is important for the lookout, who must be able to work at high elevations. Good eyesight is also important, given that the lookout must be able to see for miles, watching constantly for smoke. When the lookout spots smoke, he or she must have a clear and understandable voice to communicate with dispatchers. Lookouts must also be able to accurately read maps and be familiar with the surrounding topography in order to precisely locate a fire. Lookouts must be trustworthy, quick and steadfast. The forest service and state agencies provided training for the lookouts.

And lookouts didn't only watch for fires. They also kept weather reports, made repairs to equipment and kept adequate stock, and some even assisted in the nation's defense. Some fire lookouts were assigned additional duty as "Enemy Aircraft Spotters" during wartime. The Aircraft Warning System was established during World War II, operating from mid-1941 to mid-1944. During this time, in addition to watching for fires, lookouts also watched for enemy planes. The Aircraft Warning System was especially prevalent on the West Coast, but there were also locations on the East Coast. A few lookouts in the Blue Ridge, as far west as Tennessee, took on defense duties.

In South Carolina, where the fire tower property was about ten acres in size, the lookout also managed gardens and livestock. The size of the property allowed tower operators to grow their own vegetables and feed for their domestic livestock. In the early days of South Carolina lookouts, most tower families had a milk cow and chickens and raised a few hogs for their own use.

During the early era of fire lookouts, private companies and landowners hired men and women to work on the towers during the dry season. There were both paid and volunteer lookouts, and that continues to be the case in the twenty-first century.

By the 1930s, the Great Depression had changed the makeup of the forest service. Thousands of members of the CCC joined the forest service, building towers, manning the lookouts and fighting the fires on the ground. From 1930 to the early 1940s, a majority of the people working on the towers were young men who were previously unemployed and needed to feed their families. Most of the young men who worked for the CCC in the Appalachian Mountains lived in the area and were familiar with the land, and this made jobs as lookouts easier.

By the time World War II started, much of the manpower used by the CCC had been transferred to the nation's armed forces. Lookouts of the 1930s went to war zones in Europe and Asia, resulting in a need for new individuals to man the towers. The CCC had ceased to exist by 1942, and federal agencies began hiring new lookouts. While there continued to be some young men working as lookouts, there were also many single women, married couples and retired individuals.

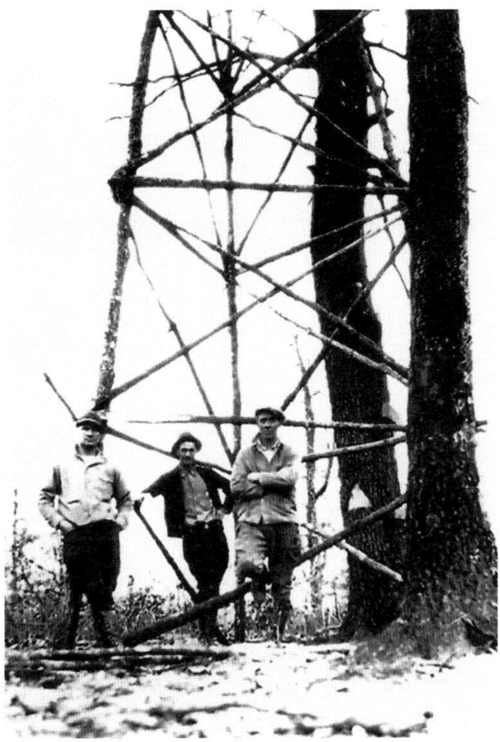

Federal employees pose in front of a makeshift fire lookout tower in western North Carolina. The tower is constructed of wood from the mountaintop. *Betty Hodge.*

Through the 1970s, federal agencies continued to hire lookouts to staff the towers. But fewer lookouts were needed by the end of the 1970s, as fire detection moved to the sky and public assistance.

In the Blue Ridge, a few fire towers remained active in the early twenty-first century, but the towers were only staffed during the driest conditions. The federal government and state forest departments continue to have a few leftover lookout staff. More lookouts in the western United States remain active than in the eastern states.

Orval Banks, a canine handler in Clyde, North Carolina, continues to man the fire tower on Chambers Mountain, one of the state's only remaining lookouts. Banks lives in the lookout cabin and regularly guides visitors to the tower. Most lookouts in the eastern United States man the towers during especially dry periods and typically have other duties in the state or national forest service.

The forest service has always tried to stay local when hiring staff to man the lookouts. Local residents were also hired because they were familiar with the landscape and could easily describe the location of fires. But during the early twentieth century, the forest service also believed that local faces would be a deterrent for arsonists and angry landowners who might have been setting fires to spite the federal government. The federal government also hired local residents for communications purposes, resulting in telephones being installed in many rural communities for the first time.

Lookouts were trustworthy, self-efficient people, but like in every profession, there were some bad apples. Some historians say that job fires became an issue in the Appalachian Mountains. As residents suffered through economic times, they would set fires in the forests, hoping that a need would arise to hire fire lookouts or firefighters. The forest service would then hire them to watch for fires, unsuspecting that the individuals who were hired actually set the fires for which they were watching. It's not known if the arsonists were successful at gaining employment, but it is possible that the forest service unknowingly hired arsonists either to fight fires or work as lookouts.

Similarly, North Carolina anthropologist William Jurgelski wrote that after World War II, a former Pisgah National Forest service employee said that the service knowingly hired arsonists believing that they would not be able to set fires due to their employment. They thought that if arsonists had jobs, they wouldn't have time to set fires.

The remote setting of the fire tower required lookouts to take long journeys to get to work, return home or just to resupply the tower. Although most of the towers were reachable by four-wheel-drive vehicles, there were a handful of towers that could only be reached by walking. In Greene County, Tennessee, a Camp Creek woman said that lookouts in her family had to hike to work at the Camp Creek Bald Tower, just above the farming community; the Rich Mountain Tower, across the state line in North Carolina; and the Meadow

Creek Fire Tower, several miles away in Cocke County. They would be gone for several days at a time, she said.

Rondall Ellis of Elizabethton, Tennessee, worked for the forest service and manned towers in the Cherokee National Forest. Ellis, who joined the forest service in 1962, recalled working on Holston High Knob Tower for seven fire seasons. "Back then, there were towers scattered all over," Ellis said. "There was one in Shady Valley, one on White Rock, near Roan Mountain, and Unicoi."

Ellis, who enjoyed the solitude of working on one of the highest towers in the Blue Ridge, said, "At that time, if one [person] saw smoke and somebody else saw smoke, they then took a cross bearing and that's where your smoke was. The tower acted like a transit."

An Osborne fire finder was mounted in the center of the Holston tower. Ellis described the fire finder as a unique piece of equipment used to pinpoint exactly where smoke was coming from. At least two fire towers would usually be used to locate fires, he said. "Then they went to a fire dispatcher, and he plotted all of the bearings and this and that," Ellis said. "At one time, Holston was the dispatching tower. All of the other towers would call in their bearings to Holston, and he would plot out where the fire was and call in the crews."

Individuals manned the tower on Holston during the spring and fall dry seasons. "It wasn't a place that you stayed constantly," he said. "You might have had rain for a few days, and you wouldn't have to stay. You could ride back home." On one occasion, Ellis said that he was on the tower for twenty-one continuous days. "We had a bad dry season," he said.

Holston's lookout is a tall steel tower topped by a small cab that housed, in addition to the fire finder, a small apartment-type stove, a refrigerator and a fold-up bed. This was a live-in tower, which meant that there were no cabins on the ground where Ellis could sleep.

"It depended on how dry it was if you stayed at night," Ellis said. "You were assigned to that tower; it was yours. If there was an emergency or something, they would send up someone to relieve you. Other than that, it was yours to take care of. It was your responsibility if you stayed at night. Usually we would stay until after dark."

Ellis said that the lookout regularly conducted his own maintenance, but for major projects, the forest service would bring in staff to do the work. During the summer months, when he wasn't working on the tower, Ellis said that he worked in the area's recreation areas and conducted timber marking, which was used to decide what trees to cut at harvest time. But when it came to fire season, the towers became the top priority.

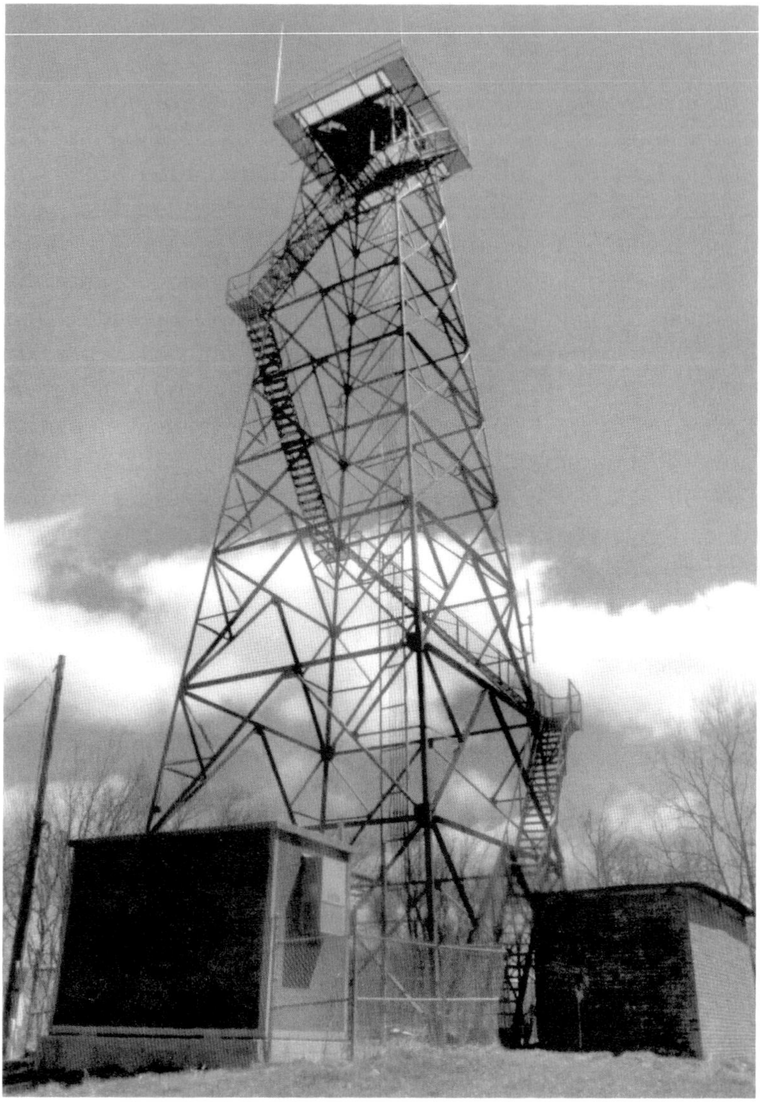

The Holston High Knob Tower was originally built for the flatlands of Mississippi. It was moved to the Blue Ridge of Tennessee in the 1940s. *Photo by Robert Sorrell.*

From atop the tower, Ellis said that on a nice clear day he could see Cumberland Gap at the Tennessee, Kentucky and Virginia border. The tower on Holston Mountain is taller than most Appalachian lookouts. In fact, while Holston's tower is an imposing one-hundred-foot-tall lookout, most others in the region are less than half its size.

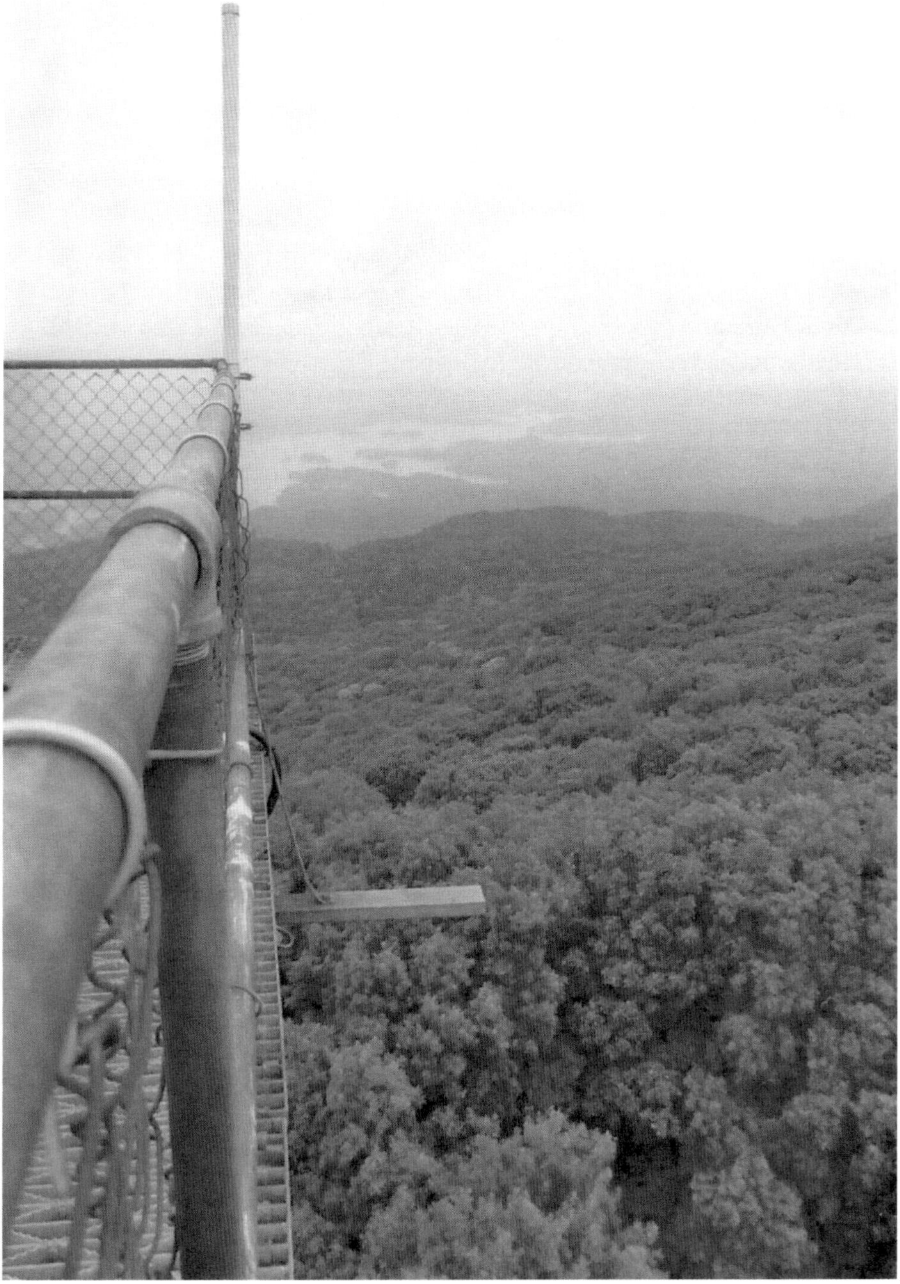

South Holston Lake can be seen from the observation deck on the Holston High Knob Fire Tower in Carter County, Tennessee. *Lisa Germaine.*

The Holston High Knob Tower is a transplant from the flat piney woods of Mississippi, where eighty- and one-hundred-foot towers are more common. Tall towers are also common along the Carolina and Georgia coasts, but not in the mountains.

Ellis said that he believed the goal of the one-hundred-foot Holston tower was to see all the way to Shady Valley in Johnson County. But Ellis said that he could not see Shady Valley when he was on the tower. However, he could see the nearby towns of Bristol, Elizabethton and Johnson City.

Ellis retired from the forest service in 1994, and he believes that the last person to work on the tower was Buddy Phillips. "There are not many of us left," he said. His friend, Gale Colbaugh, also has many memories of the tower on Holston Mountain. His father, Luther Colbaugh, worked on two towers that stood on top of Holston High Knob and witnessed more than just the forest catch fire.

Gale Colbaugh recalled that his father was on the old wooden tower on October 9, 1943, when it caught fire. The old tower, built in 1938, was sixty-five feet tall and featured a fourteen- by fourteen-foot cab. Luther Colbaugh, the tower's primary lookout, was a member of the CCC when the tower was completed. "[Luther Colbaugh] has a reputation of being a dependable, trustworthy employee," noted an inquiry report into the fire.

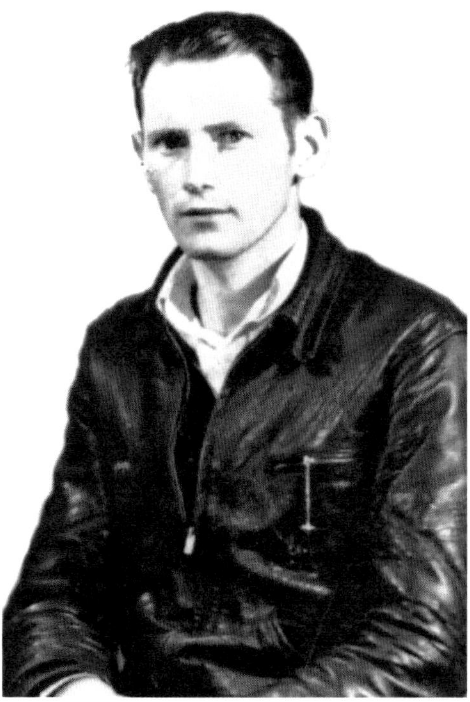

Luther Colbaugh, a U.S. Forest Service employee, worked on two fire lookout towers on Holston High Knob in Carter County, Tennessee. *Gale Colbaugh.*

Gale lived in nearby Liberty Hollow and could see the tower from his front yard. When everything on the tower was satisfactory, his father would light an old Aladdin oil lamp. The light of the lamp, he said, could be seen from the home in Liberty Hollow. "If it was lit, we knew he was okay," Gale said.

The Aladdin lamp was actually a common accessory of the nation's lookout because it was not considered a fire hazard and provided sufficient light. Battery-operated light and electric lights became more desirable, however, as they became available.

The light that came from the tower on October 9, however, was not from the lamp. Gale said that his father, Luther, was manning the wooden tower the day it caught fire. Luther's fourteen-year-old brother, Herbert Colbaugh, was also on the lookout.

According to the inquiry, Luther and his brother had hiked up the mountain, carrying supplies with the intent of occupying the tower. They arrived about 11:30 a.m., the report said, and they completed installing the haul-in cord on the "lazy-gal" line, which was used to supply water to the cab from the spring; worked on the battery charger; stored supplies; got up water for kitchen use; and cleaned up the quarters.

At about 4:00 p.m., they started a fire in the stove, using oak and/or chestnut wood, to heat water for scalding the dishes. The report said that they used splinters from the wood to start the stove fire, using no kerosene or paper. In addition, the report stated that it did not appear that they made an exceedingly hot fire. After heating the water, the fire was allowed to die down.

At about 4:30 p.m., the report said, or possibly later, a heavy electrical storm occurred. Lightning cracked around close and rang the telephone bell. The wind was blowing intensely, from six to ten miles per hour. A few minutes after 5:00 p.m., Luther Colbaugh told investigators that he smelled smoke and saw smoke coming down past the windows on the outside. Since the fire in the stove had died down, he suspected that a fire had ignited and opened up the twelve- to fourteen-inch square scuttle in the ceiling. That's when he observed that a hole about fifteen inches across had burned in the roof.

The lookout was unable to climb up into the scuttle, so he threw a dishpan of water up through the hole toward the fire. He then retrieved a fire extinguisher. After exhausting the one-quart fire extinguisher, he stepped outside the cab, stood on the railing and threw about three gallons of water on the fire. The report said that these efforts cooled the fire off, but the downdraft of the wind over the roof fanned the flames up again.

Luther Colbaugh then called the ranger's office for help. Two men rushed to the scene and arrived with two five-gallon backpack pumps at 6:00 p.m. After calling the ranger's office, Colbaugh then started drawing water from the spring, which took about seven minutes to do. He stood on the rail again and threw the water on the roof. He and his brother eventually climbed down the tower and escaped from the fire to safety. "It was terrifying knowing he was on the tower," Gale Colbaugh said.

Gale said that his family could see the flames on the mountain, and they feared that Luther was still in the cab. Fortunately, both individuals made it to safety without any injuries, but the tower was a total loss.

Following the fire, the National Forest Service conducted a full inquiry into the cause. According to the inquiry, the fire was likely caused by either a spark from the stove pipe or lightning. The inquiry also said that if the tower had the appropriately sized roof and ceiling scuttles, "Colbaugh would have been able to more effectively fight the fire and may have been able to control it with the means available."

Soon after the fire, the forest service purchased a replacement one-hundred-foot steel tower from Mississippi and placed it on top of the Holston High Knob. Luther Colbaugh became the first person to man the new steel tower. He worked on the tower during the dry season.

Gale said that when he was young, he would ride the school bus to the head of Rome Hollow on Friday nights to visit his father. He would follow the telephone lines, which were installed along trails, to reach the tower from Rome Hollow. He would then stay Friday and Saturday night and return home Sunday. The father-and-son duo would work on the tower, maintain the property, cook meals and spend quality time together on top of the mountain.

In 1955, Luther Colbaugh was transferred to Greeneville, where he oversaw fire control for Greene and Cocke Counties. "He was transferred there to help curb forestry fires and improve on community relations," Gale said. After his father went on to Greene County to work for the forest service, Gale said that he saved that old oil lamp that his father had lit on the tower. He also has several other items that his father used to mark timber and map new roads.

Luther Colbaugh retired after thirty-eight years of service due to medical conditions. "He was truly dedicated to his work, and it was the delight of his life," Gale said.

Only a handful of individuals worked atop the tower on Holston Mountain. L.A. Peters also worked on Holston High Knob. His son, Keith Peters, who lives in Texas, recalled that his father worked for the fire service department and was responsible for installing electric power to the tower. He installed the wires to all the poles going up to the tower. "He also made a zip line from the top of the tower to their water source so he could get water to the top of the tower," he said. "My father only went to the seventh grade, but he was the smartest man I ever knew."

Jack Kestner, a newspaper reporter from Washington County, Virginia, spent one season in a fire tower on Clinch Mountain in the Channels

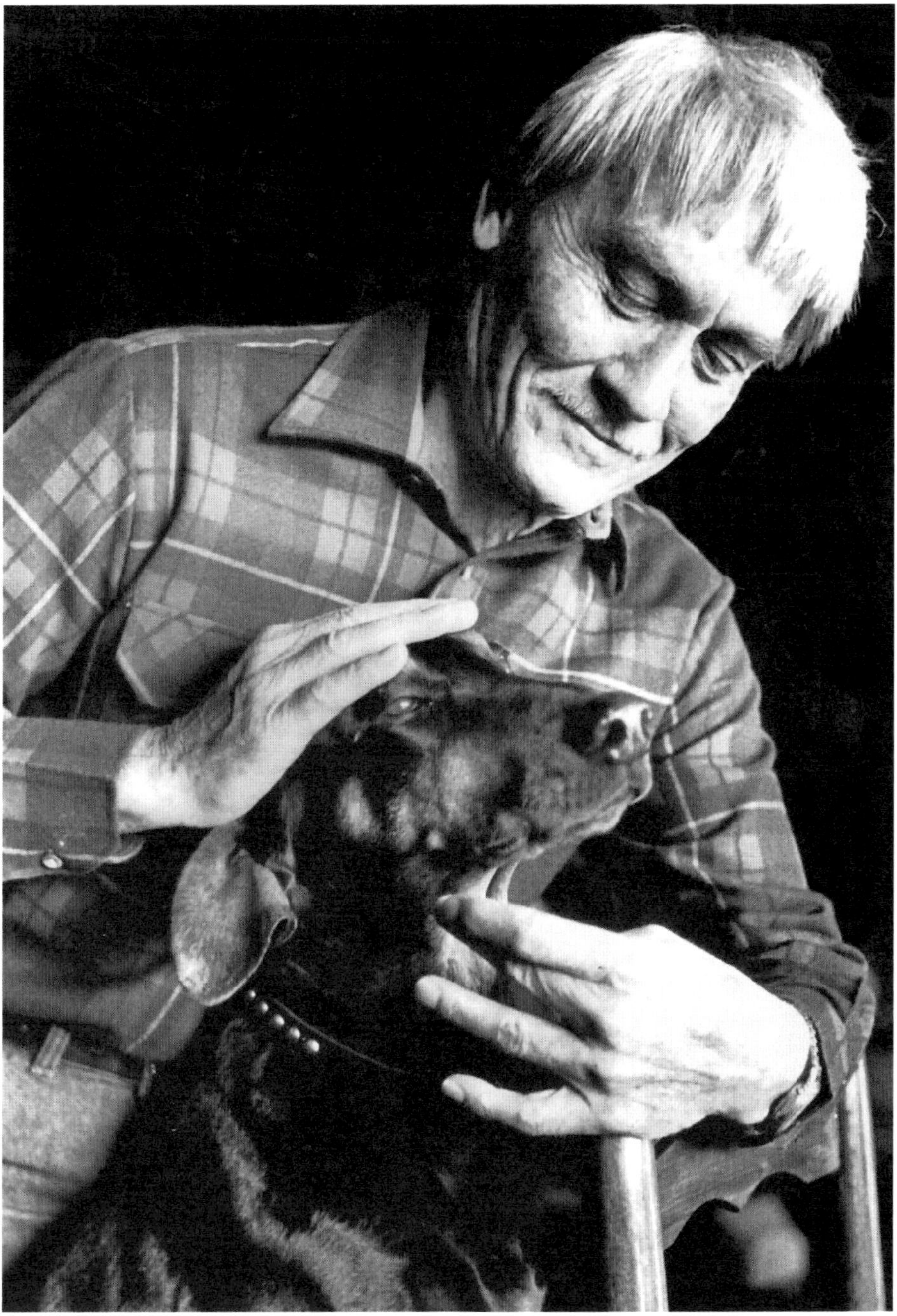

Newspaperman Jack Kestner, who grew up in Hayters Gap, Virginia, worked briefly on the community's fire tower. *Earl Neikerk.*

State Forest. He had grown up in the Hayters Gap community of Washington County, where he was born in 1921. His newspaper career took him to the *Ledger-Star* in Norfolk, Virginia, where he worked as a military reporter.

His daughter, Lisa Kestner Quigley, who now lives at the family homestead, just below the lookout tower, said that her father enjoyed hiking to the Hayters Knob Fire Tower when he was a teenager. The lookout is also known as the Middle Knob Fire Tower.

After Kestner returned to Washington County, he began writing a column for the *Bristol Herald Courier*. In 2006, a selection of Kestner's columns was published in the book *A View from the Mountain*. A few of his columns were about his stint on the tower, which was originally built by the CCC. On April 19, 2001, he wrote, "44 years ago today during a break in my journalistic career, I'd gone to work as a fire tower lookout at Hayters Knob. I'd been up there a couple of weeks and was running low on supplies when I recalled that my father had mentioned he'd like to visit the tower."

Whenever Kestner needed supplies, needed to go into town or had an emergency, he would radio District Headquarters in Abingdon. On this occasion, he asked that his father be telephoned and given a grocery list. "A few minutes later, the radio crackled to life with the news that Dad would be coming up the following day, most likely arriving in early afternoon," Kestner wrote. "By coincidence, the following day was his 70th birthday."

Kestner said that his father hadn't arrived by 4:00 p.m., so he radioed headquarters to ask if he had left nearby Bristol. They radioed back to say that his father had left around noon, which meant he was about two hours late. The columnist said that the road from Route 80 ended at a clearing near the beginning of the switchbacks. This trail zigzagged up the side of the mountain. In the 1930s, Kestner said, it was the route used by mule teams carrying metal components of the tower. "One can only pity those poor mules," he wrote.

Kestner said that he knew his father's car would have a difficult time navigating the route, so he began walking down the switchbacks. He found the car parked at the clearing, which meant that his father had wandered off the trail. Kestner found some footprints and began to follow them, hoping to find his father. He eventually noticed that his father had gone off the trail.

After finding some groceries in the brush, Kestner tried to call for his father but received no answer. So, he hurried back to the tower and radioed for help. Headquarters began alerting people by phone. Around midnight, he said, his father made his way off the mountain and to a telephone. "And, boy, did he have a story to tell," Kestner wrote. "He said that when he suddenly found himself 'lost,' he'd decided to follow a stream bed down to the valley."

The columnist said that lookout duty in the seven- by seven-foot cab atop the one-hundred-foot tower "fulfilled one of my childhood dreams of entering the forestry service. I loved the great outdoors. This ambition was eventually replaced by a dream of writing books. Little did I realize at the time that one day they'd be linked."

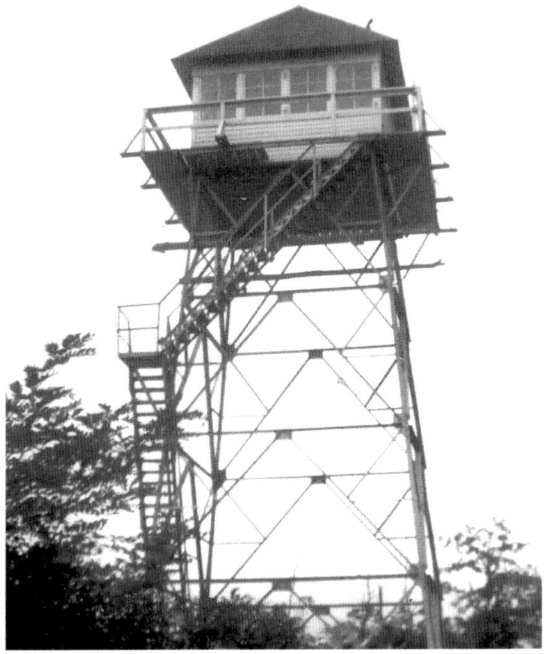

The Hayters Knob Fire Tower and small live-in cabin were constructed in the 1930s in Washington County, Virginia. *Lisa Kestner Quigley.*

Kestner also utilized his experience as a lookout to write the book *Fire Tower*, a juvenile adventure novel published by Funk & Wagnalls in 1960. For the book, Kestner said that he borrowed heavily from his experiences. On his first night at the tower, he said, he woke up to find that the door of the little one-room shack had blown open and his blankets were covered with snow. "When I shined a flashlight out into the darkness, I discovered that clouds had descended on the mountaintop, clouds filled with swirling snow," he wrote.

Kestner also recalled an incident that likely ended his lookout career. "Driving out for supplies in my World War II jeep early one morning, I ran over a ledge and the jolt sent my back into spasm," Kestner wrote in a column. "A total lock-up. In excruciating pain, I opened the door and fell to the ground. And there I lay. Helpless." In the late afternoon, Kestner said that he heard an approaching thunderstorm. "Realizing that if I got soaked I'd probably suffer hypothermia," he wrote, "I reached down and cut a lace from my boot, tied a couple of sticks together to form a hook, and used it to pull the jeep door over until I could reach it."

He then pulled himself to the jeep and fell inside. A search party, which included District Fire Chief Ray Duncan, then found him around midnight. His daughter believes the search party saved his life.

Kestner noted in his column in 2001 that the era of manning towers had passed, replaced by heightened communications. "I think is a shame, for it was a job I dearly loved," Kestner wrote. "Some would probably consider it boring. There really wasn't much to do except keep an eye out for smoke in all directions. If one showed up, a bearing was taken on it with an alidade mounted on a map table occupying the center of the cab. The bearing was then radioed to district headquarters."

The former lookout said that at headquarters, the duty man would stretch a string from the tower's position on a huge wall map. The duty man would then request a bearing from another nearby tower. In Hayters Knob's case, the duty man would check with towers at Mendota (Washington County), High Knob (Wise County) or Buckner's Ridge (Scott County). Wherever the strings crossed was where the fire was located.

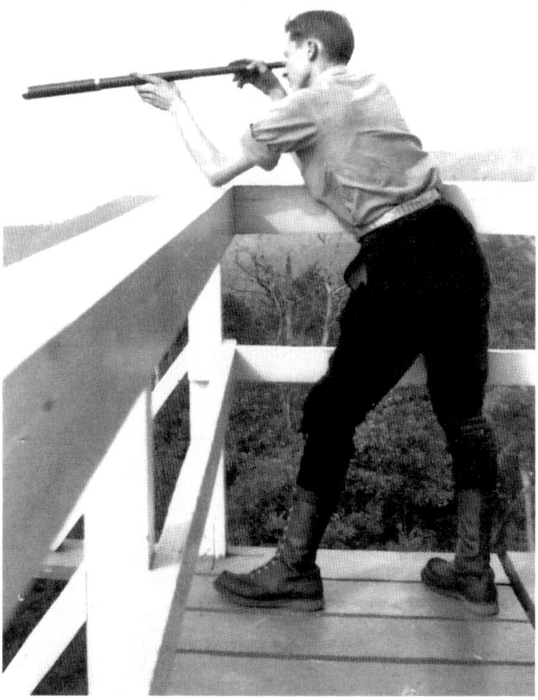

A man uses a monocular to watch for smoke and flames from the historic Hayters Knob Fire Tower in the Channels State Forest of Virginia. *Lisa Kestner Quigley.*

The tower itself was pretty bare, Kestner's daughter said. She recalled seeing a cot, a little desk and a stool. There was no electricity, and water was pumped to the tower. "You couldn't have company up there," she said. "This was such a little place. There was a cot and a heat-up stove. He loved it because he could sit there and just look outside."

"One of the things I enjoyed most was watching spring slowly climb the south side of Clinch Mountain from the 1,500-foot valley floor to the tower's 4,000-foot peaks," Kestner wrote. "It was an advance marked by blooming dogwood and redbud and serviceberry, the leafing out of trees. Spring creeping upward in slow increments."

In another column, Kestner wrote, "I watched the first greenery of spring show up down in the valley, then move in slow day-by-day steps up the side of the mountain. I watched sunrises and sunsets, and circling hawks, and rainbows I could actually look down on. It was only incidentally that I watched for smoke—not once imagining that in another 20 years I'd build a retirement home on the side of the mountain below the tower."

Kestner built a home on the side of the mountain near the tower. His family still lives in the home. The tower has become somewhat dilapidated, and the small one-room cabin where he slept is in shambles. The state has discussed dismantling the tower, but Kestner's family has fought it.

Colleen Taylor was the last person manning Hayters Knob when it was decommissioned about 1970. Like many American industries, the forest service hired women during wartime. Starting in World War I, women were hired to staff the nation's fire lookouts. Hallie M. Dagget became the first female fire lookout in the forest service in 1913. She spent fifteen years working at the Eddy Gulch fire tower in California's Klamath National Forest.

The lookouts on Hayters Knob worked closely with the lookouts on the nearby Mendota Fire Tower. In a 1951 issue of *Boys Life* magazine, there is a story about twelve boy scouts from Troop 16 of Bristol visiting Mendota. At the time, Gene Olson was the lookout. The magazine story revealed that the boys had camped at nearby Moccasin Valley. At the time, the road to the Mendota tower was passable, and Olson drove the boys up the mountain in his pickup truck.

While the boys had lots of questions about forest fire detection, the article's writer said that Olson preferred to show the scouts firsthand. They all climbed to the top of the tower, where Olson communicated with another distant lookout to pinpoint the scout's campsite in Moccasin Valley. "When Olson saw their surprise at the precise way they could locate their camp, he simply explained. 'You gotta be accurate when you're dealing with forest fires,'" the article stated.

Olson, a regional forester, also worked with the boys to fight an imaginary fire. They constructed fire lines and utilized an Indian Fire Pump, which firefighters have used to battle wildfires for nearly a century. Firefighters carry the fire pump on their backs and can easily adjust the flow of water.

Although fire lookouts could spend days on the towers without seeing a soul, there were occasional visitors. Judy Honeycutt recalled her parents

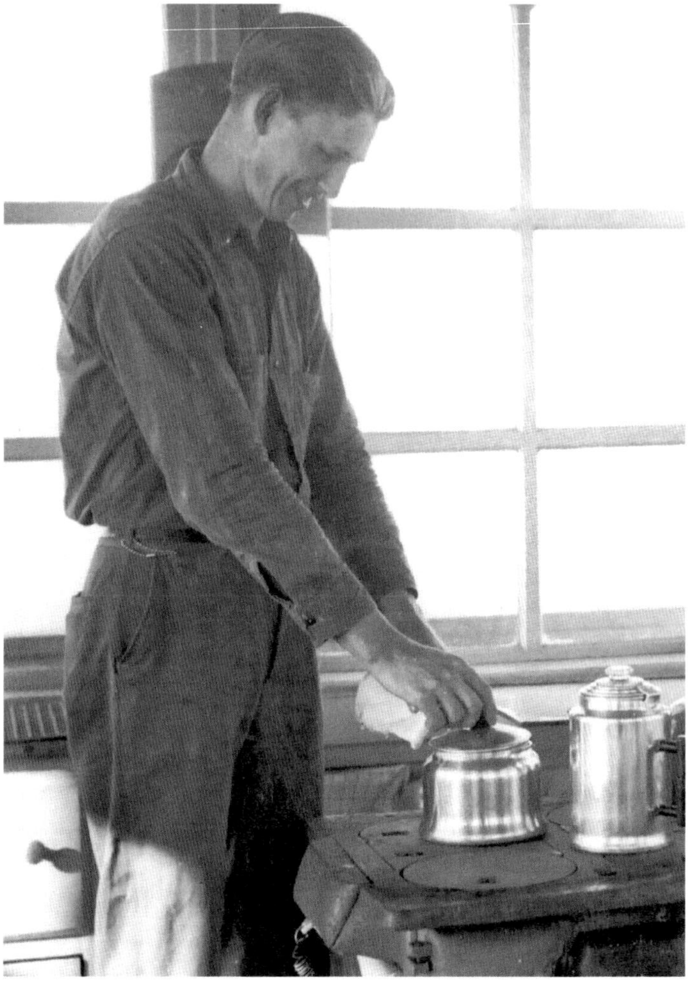

Fire lookout A.B. Patton prepared his meals on a wood-burning stove in the cab of the Cold Spring Fire Tower in the Cherokee National Forest of Tennessee. *National Archives.*

visiting Earl Haun on the Pinnacle Mountain Fire Tower in Tennessee in 1935. They hiked up the mountain after their wedding and spent some time with Haun, who cooked a meal for them.

Fire lookouts, especially along the Appalachian Trail, routinely encountered hikers stopping for a view. Lookouts at Albert Mountain, Rich Mountain and Wayah Bald regularly provided hikers with tours of the towers and demonstrated fire detection techniques.

CHAPTER 4

END OF AN ERA

Early twentieth-century forest fire policy resulted in the creation of a vast network of fire lookout towers, so wildfires could be spotted and extinguished as quickly as possible. Forest Chief Ferdinand Silcox mandated that fires were to be controlled by 10:00 a.m. the day after they were first spotted, which led state governments and the federal government to build thousands of towers across the country, from the coastal lowlands to the mountain highlands.

But by the mid- to late twentieth century, foresters had begun to reconsider the importance of fires. Could natural fires be needed to control growth and life in the nation's forests? Silcox's policy created a heavy amount of brush and dead plants in the forests that would have been naturally burned off by fire if foresters had not suppressed every blaze. By the 1970s, the forest service and park service had a change of heart and moved to a policy of prescribed fires to achieve resource management. The "10:00 a.m. policy," which Silcox ordered in 1935, was rescinded in the late 1970s.

During the late twentieth century, foresters used fires to burn off brush, undergrowth and dying forests. If fires got out of hand, firefighters would quickly get them under control before they encroached on structures and live vegetation. Foresters now say prescribed fires reduce fuels for larger wildfires. The change in fire management resulted in a decrease in the number of manned lookouts.

There has been some debate during the early twenty-first century, however, as fires have grown larger and larger, devastating millions of acres of forestland throughout the western United States. As a result, lookouts continue to be used in California, Oregon and a few other states.

In addition to a policy of controlled fires, new technology has also affected the use of fire lookouts. The U.S. Forest Service began using aircraft around the country in the 1950s to detect and fight wildfires, although planes were used as early as 1917 in California. And although the technology has not been guaranteed to spot every wildfire, satellites began monitoring the nation's forests in the late twentieth century. As the technology improves, satellites will likely be used more often for fire detection in the twenty-first century.

By the late 1990s, cellular phone usage had dramatically decreased the need for fire lookouts. The public began calling in wildfires on their cellphones. Dispatchers usually receive multiple calls about fires. Wildfire calls are similar to traffic accident 911 calls. Emergency communications officials say that when there is a traffic accident on a major thoroughfare, they receive multiple phone calls from passersby. So when smoke billows from the forests and is visible in nearby communities, 911 dispatchers typically receive numerous phone calls from concerned citizens. But as many cellphone users know, service is not always available in the mountains, so those extremely remote fires cannot be called in to 911 dispatchers. For that reason, cellphone companies have begun placing towers in the forests. Cellphone towers have even been mounted on old decommissioned fire towers.

By the twenty-first century, agencies had begun using surveillance cameras and drones to detect wildfires. Aircraft and satellites also continue to be used.

Starting in the 1960s, the U.S. Forest Service, National Park Service, Bureau of Land Management, Bureau of Indian Affairs and state forest departments began decommissioning fire towers, as most were no longer needed to detect fires. Between 1970 and 1990, hundreds of fire lookout towers were decommissioned and dismantled.

There are a handful of towers in the Blue Ridge that continue to be manned during exceptionally dry periods, but there are far fewer than the number of towers still used in the western United States.

The Great Smoky Mountains National Park, which records about two lightning-ignited wildfires every year, still has four fire lookouts, down from sixteen towers during their heyday. When the park was established, national parks across the country had a policy to suppress all fires. To fight wildfires, which were usually reported in late spring, when strong and severe thunderstorms passed through the mountains, the park service in the 1930s developed a management plan that required suppression of all fires. The

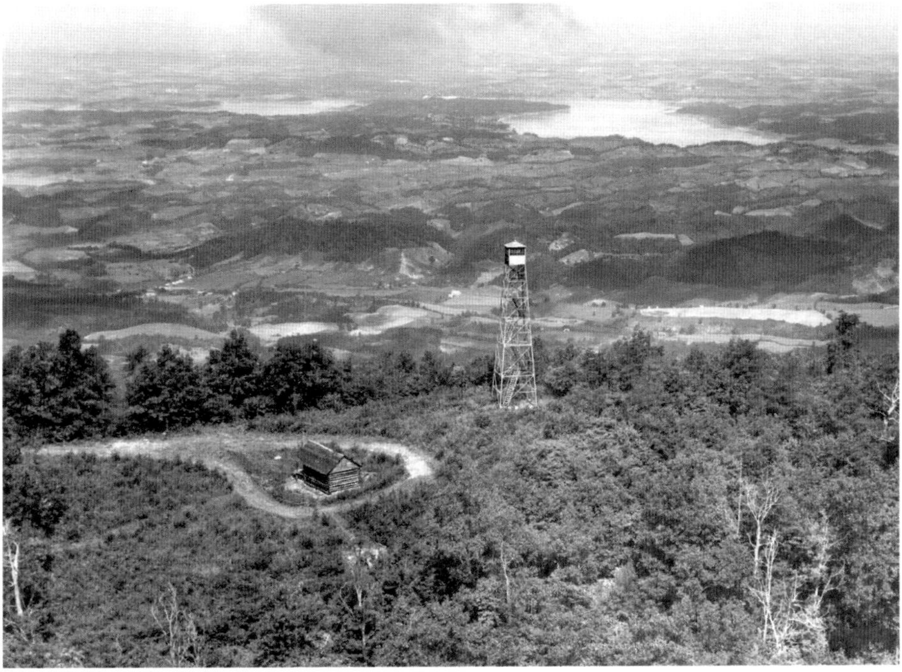

The English Mountain Fire Tower, located on one of east Tennessee's highest mountains, featured a small live-in cabin. *Tennessee State Library and Archives.*

park service, with assistance from New Deal–era agencies such as the CCC and WPA, constructed sixteen lookouts throughout the region, including outside the park's boundaries. Debris and undergrowth were also cleared, and firebreaks were created.

The park utilized lookouts at Bunker Hill, Rich Mountain, Blanket Mountain, Cove Mountain, Greenbrier Pinnacle and Mount Cammerer, also known as White Rock, in the northern section of the Smokies. The southern section of the Smokies included Shuckstack, High Rocks, Spruce Mountain and Mount Sterling. Through cooperative efforts with external agencies, the park service also utilized Look Rock, Hall Top, English Mountain, Frye Mountain, Mount Noble, Barnett Knob and Sutton Mountain.

Lookout towers at Barnett Knob and Mount Noble were managed by the Bureau of Indian Affairs, but in coordination with the park service, it assisted in fire detection in the national park.

The park slowly began decommissioning fire towers starting in the 1970s. By the mid-1980s, the park's administration had decided that the abandoned fire towers were a liability and should be removed from the park. Removing the towers became a challenge. Heavy machinery was not allowed to access the remote locations. Eventually, the Tennessee Valley Authority flew helicopters over the sites, where ground crews had bundled the structures. Staff and supplies were lowered to the ground from the helicopters or, in many cases, hiked in by foot or pack animals. The helicopters transported the towers to designated locations around the park, and then they were trucked away.

All but four towers within the national park were removed. Lookouts at Mount Cammerer, Shuckstack, Cove Mountain and Mount Sterling remain standing in the park. Several cooperative towers, including English Mountain and Barnett Knob, also still stand.

One of the removed towers, at Blanket Mountain, was a steel structure built in 1934 by the CCC. It had replaced a wooden tower constructed in 1927 by Tennessee's forestry department. The steel tower was sixty feet tall. The CCC also built a new lookout cabin at the site. After the cabin and tower were removed, hikers could still find pieces of the structures in the ground. The site is located along a popular hiking trail that connects to the Jake's Creek trail near Elkmont.

At Cove Mountain, the park service uses the old sixty-foot-high fire tower to monitor air quality. The steel Aermotor tower, built in 1935, is not accessible to hikers. An equipment shed has been built into a section of the tower, blocking the stairs. Mount Cammerer, Shuckstack and Mount Sterling are accessible to hikers, but the forest service only recommends access to Mount Cammerer, a ground-level cab.

State governments and the national forest service also have dismantled dozens of fire lookouts in the Blue Ridge. At one time, the Cherokee National Forest in Tennessee had eighteen towers. Most of the towers have been removed, including the White Rock Tower in Carter County and Buck Ridge Tower in Johnson County.

The Forest Fire Lookout Association, an organization dedicated to the preservation of lookouts, has attempted to keep track of fire towers in the Blue Ridge and throughout the country. According to the organization, as of late 2014, 40 towers remain of the original 250 lookouts in Tennessee. Georgia, which had 360 towers, now has 172. In North Carolina, there were 210 towers, and in South Carolina, there were 179 towers. In 2014, both North Carolina and South Carolina have 60 towers remaining. Virginia had a peak of 197 lookouts. There were 50 towers left in 2014.

While dozens of fire towers remain standing throughout the Blue Ridge Mountains, they have been abandoned and are in poor dilapidated condition. Very few towers—including the High Knob Tower in Virginia, Pinnacle Mountain Tower in Tennessee and Mount Cammerer in the Great Smoky Mountains National Park—have been restored for visitors.

Towers remain standing in 2014 at Holston Mountain, Bays Mountain, Chimney Top, Meadow Creek, Camp Creek Bald, Little War, Bulls Gap, English Mountain, Kettlefoot, Sharps Ridge, Lenoir City, McMinn Ridge, Starr Mountain, Madisonville, Ocoee and Millstone Gap in east Tennessee, and this does not include remaining towers in the national park.

Tennessee is also home to the Clingmans Dome and Look Rock observation towers, which previously were fire lookout towers. The original towers have been torn down and replaced with concrete structures.

Remaining western North Carolina towers outside the national park include Albert Mountain, Barnett Knob, Bearwallow Mountain, Big Knob, Boone, Chambers Mountain, Cowee Bald, Duckett Top, Fryingpan Mountain, Green Knob, Hawshaw Mountain, Joanna Bald, Little Snowball, Locust Knob, Moores Knob, Mount Noble, Panther Top, Parrydise (also known as Little Scaly Mountain), Rendezvous Mountain, Rich Mountain, Toxaway Mountain, Walker Top, Wayah Bald, Wesser Bald, Woody's Knob and Yellow Mountain.

The State of North Carolina continues to man two of its fire towers at Chambers Mountain and Toxaway Mountain. The North Carolina Forest Service was unable to provide any information about the two lookouts, but it noted that they are occasionally used for fire detection. Like other states in the Blue Ridge, North Carolina also uses other towers for communications. The U.S. Forest Service in North Carolina has set aside thousands of dollars for fire tower restoration in recent years.

A few fire towers still stand in the Blue Ridge of South Carolina's Upstate, including Issaqueena, Long Creek, Long Mountain and Paris Mountain. Several towers still stand in the state's lowlands between Columbia and Charleston, where some of South Carolina's first lookouts were constructed.

Georgia's remaining Blue Ridge lookouts stand at Black Mountain, Brasstown Bald, Chenocetah, Fort Mountain, Glassy Mountain, Grassy Mountain and Rabun Bald. Dozens of towers, including lookouts that are still manned, remain standing in central and southern Georgia, including several along the Georgia coast.

A 2006 article published in the *Savannah Morning News* noted that Georgia, which has had 360 lookouts, still has 7 manned towers in the state. Georgia probably mans more towers in the twenty-first century than any state east of the Rocky Mountains. In the 1960s, there were more

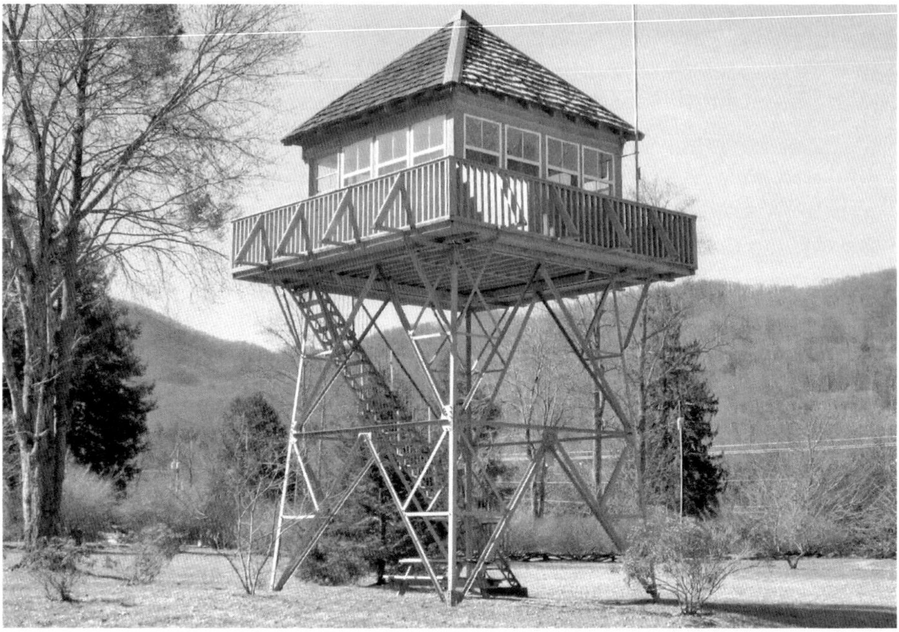

The Big Ivy Historical Park in Barnardsville, North Carolina, is anchored by the relocated Little Snowball Fire Tower. *Photo by Robert Sorrell.*

than 200 towers manned in the state. The state and local governments split the cost of paying lookout salaries.

There are several fire lookout towers still standing in Virginia, including Bearwallow, Big A Mountain, East River Mountain, Hayters Knob, High Knob, High Knob–Leemaster, Hunters Gap, Keen Mountain and Mendota in the southwestern portion of the state. Just east of Interstate 77, the Groundhog Mountain Lookout is located on the Blue Ridge Parkway. Bull Mountain, Butt Mountain and Cove Mountain are nearby. Several towers also remain standing in the Shenandoah Mountains and coastal areas of Virginia.

In 2014, Ed Stoots, regional forester for the Virginia Department of Forestry, said that the state no longer operates any towers other than a few that have radio repeater equipment mounted using the tower as a base for antennas or other equipment. For all others, the state has either disposed of them to the local landowner or, in the case of Hayters Knob, the state still owns it because it's on state forest property. Stoots noted that the bottom steps of Hayters Knob have been removed to keep people from climbing the dilapidated structure.

The Virginia State Police has also used fire towers during the last few decades, according to spokeswoman Corinne Geller. State police mounted communications equipment on the towers, including one in Patrick County.

Southwest Virginia is also home to several unique lookout sites, including Big Walker in Wythe County, a historic observation tower, and the Birch Knob Observation Tower, a modern lookout at the site of a former fire detection lookout.

The tower on Big Walker Mountain (3,405 feet) was originally built in 1947. According to the Virginia Department of Game and Inland Fisheries, the lookout has been run continuously by the Kime family. The original tower has been rebuilt and restored, and the former chairlift leading to the top of Walker Mountain has been removed. On a clear day, some time ago, from the top of the 100-foot Big Walker Lookout Tower, visitors could see five states. Haze has deteriorated the view, but visitors can still see at least three states.

An article in the Richmond, Virginia *Times-Dispatch* in 2014 reported how Stuart Kime built the tower in 1947 on Big Walker, which straddles the Wythe and Bland county line. "Stuart Kime was a machinist, designer and entrepreneur who could build just about anything," Bill Lohmann wrote in the article. "Before World War II, he worked for a man in the Ozarks who had a tourist attraction with an observation tower. He fell in love with the business, Ron Kime said, and acquired the plans for the tower with an eye toward having one of his own someplace." The article continued: "That place turned out to be Big Walker Mountain, where Kime and his wife, Abigail, settled in 1947. They built a gas station and gift shop, and Stuart commenced work on the tower."

Just outside the Great Smoky Mountains National Park, the Eastern Band of Cherokee, the only federally recognized Native American tribe in the Blue Ridge, operated at least two fire lookouts on its land, including Barnett Knob and Mount Noble. Barnett Knob, which was built in 1932, was one of the last used fire lookouts in the eastern United States. The Bureau of Indian Affairs, which built and operated dozens of fire towers around the country, manned the tower regularly until 2004 and continues to utilize the tower during excessive dry periods. The 80-foot metal tower features a 7- by 7-foot metal cab. A wood cabin is near the tower's base. The tower remains in excellent condition and is maintained by the BIA and Cherokee tribe. Barnett Knob is a 4,665-foot peak in the Cherokee Indian Reservation and has also been used for government communications.

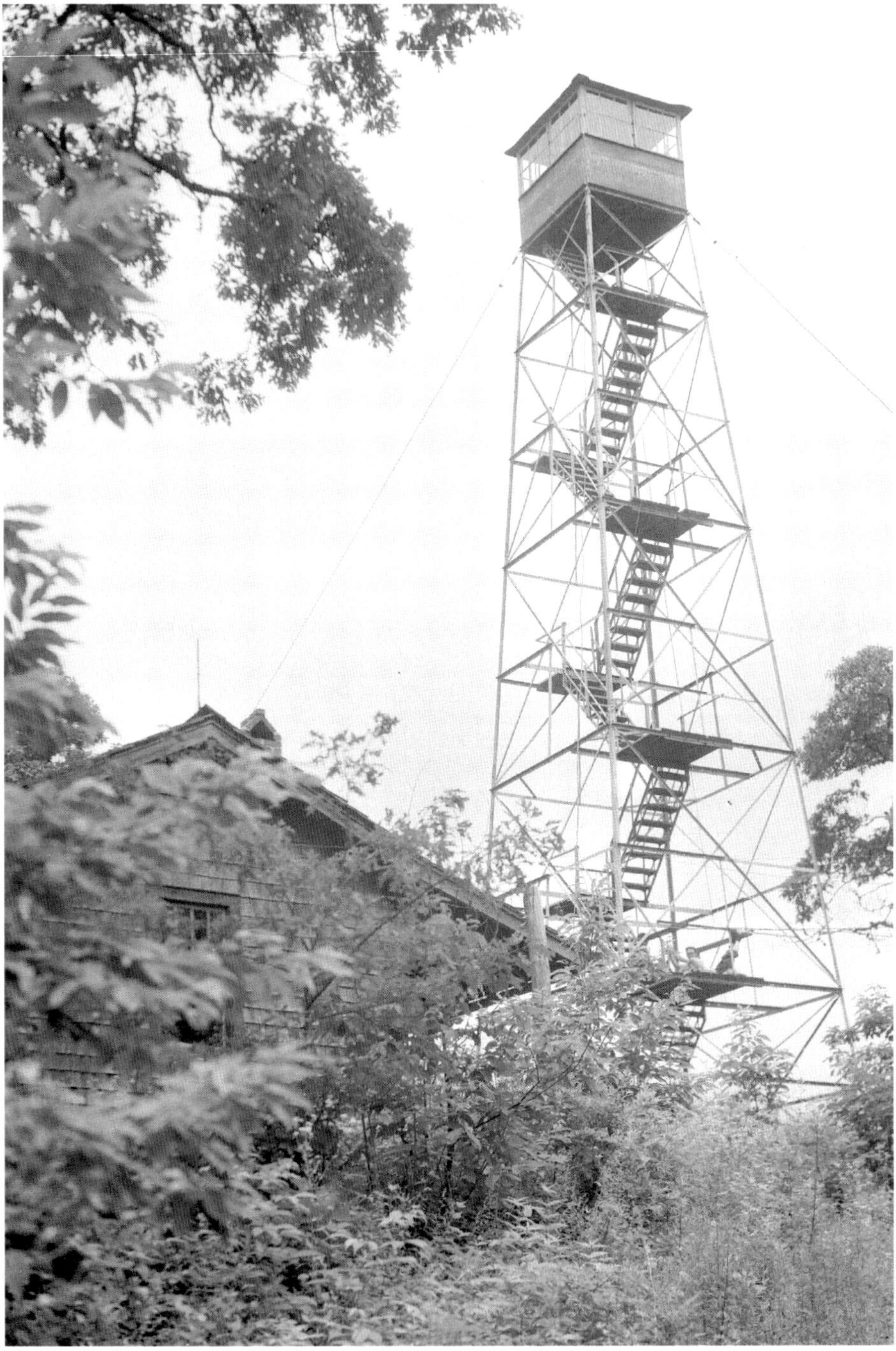

The Barnett Knob Fire Tower, managed by the Eastern Band of Cherokee and Bureau of Indian Affairs, is located in North Carolina. *Albert "Dutch" Roth Collection, University of Tennessee.*

A wood tower was originally built on Mount Noble in the 1930s and was replaced in the 1950s with a steel Aermotor tower, although it was only used for about a decade. While Barnett Knob, to the east, is still occasionally manned, the Bureau of Indian Affairs no longer utilizes Mount Noble, and it has somewhat deteriorated.

At many of the remaining towers, the lower steps have been removed to prevent visitors from climbing the structures. Other towers have been fenced in, razor wire has been installed and gates have been locked. Many of the metal towers have become rusted. Several towers also have lost steps and hand railings, making a climb up the lookout dangerous.

The old steel tower at Mendota on Clinch Mountain has become exceedingly dangerous to climb, resulting in the removal of stairs from the bottom portion of the tower. Many of the remaining stairs have broken and fallen from the tower. And vandals have covered the tower's cab and the nearby block building with graffiti. The flooring in the cab is missing.

The tower, built above the sleepy village of Mendota, has become a destination for bird watching. It has been given the nickname of the "Hawk Capital of the World," as many hawks, falcons and other birds can be seen flying around the fire tower. Although there is a passable gravel road that stretches across Clinch Mountain, allowing access from Mendota, an old jeep spur road is impassable.

Jack Kestner, who had manned Washington County, Virginia's Hayters Knob Tower for one season, wrote about the old lookout in September 1990. "Chatting with District Forester Harold Hannah the other day, I learned that Hayters Gap fire tower, located above me on top of Clinch Mountain, is being declared surplus and will probably be sold and moved," he wrote in a column published in the *Bristol Herald Courier*. "Sad news. That tower's been woven in and out of my life since I was a teenager."

Hannah told Kestner that the Virginia Division of Forestry had been spending less on towers and more on air surveillance and citizen reportage of fires. It also had become hard to find appropriate lookouts to staff the towers during the spring and fall burning seasons, particularly isolated towers like Hayters Knob.

"Vandalism and its accompanying rise in maintenance costs have also become a very real problem at these remote locations," Kestner wrote. "Towers at Mendota and Leemaster in Buchanan County are likewise being declared surplus, Hannah says, leaving only a half a dozen in the district. Of these, only three critical ones will be manned, while the rest will be retained as radio antenna towers to relay district communications."

None of the towers Kestner mentioned in 1990 has been removed, including Hayters Knob, Mendota and Leemaster. "Fire control is simply a different ball game than it was a generation ago," Hannah told Kestner. And it is.

In 2001, Kestner returned to the lookout for a visit. He said that the windows of the cab at the top of the tower had a few bullet holes. Also vandalized, he said, was the nearby one-room cabin. During that visit, Kestner could recall where he kept things when he was the lookout. In the cabin, he said he could remember where the supplies were stored, where the little wood-burning stove sat and where the iron cot rested.

Similar to the tower at Mendota, the government has removed the bottom run of stairs from Hayters Knob. In 2014, the ramshackle one-room cabin is missing walls and could seemingly collapse with strong winds. The wooden cabin is covered by a tin roof.

Although there are still fire lookout towers standing across the Blue Ridge, many other lookouts are gone. As the national park system developed plans to remove towers from the Great Smoky Mountains National Park, the state and national forests also removed towers.

Some remaining lookout towers in the Blue Ridge Mountains are now open to the public as observation lookouts, such as Wayah Bald, Mount Cammerer and Bays Mountain, near Kingsport, Tennessee. The sixty-foot-tall Bays Mountain Tower, known as the Garden Mountain Tower on topographic maps, is located within the three-thousand-acre Bays Mountain Park, a city-owned recreation area. The CCC built the steel tower on Bays Mountain in 1937, according to a sign on the lookout, in conjunction with the Tennessee Division of Forestry. From the tower's cab, visitors can see the cities of Kingsport and Johnson City, as well as the Eastman Chemical Company and the Holston Army Ammunitions Plant.

Another tower accessible to the public is located on the North Carolina and Tennessee state line at Rich Mountain. The forest service and CCC Camp Alex Jones built the thirty-foot-high steel and wood lookout tower near Hot Springs, North Carolina, in 1932. It features a fourteen- by fourteen-foot wooden cab, and given the numerous signs of etchings and graffiti, it has seen its fair share of visitors. The interior shows some damage to the walls and plywood floor. An accessible catwalk, featuring a hatch leading to the steps, wraps around the cab.

The tower was last manned in the 1990s, but the live-in cabin has not been slept in, by a lookout, since the 1970s. The tower is easily accessible by a road that loops around the tower at a small grassy clearing. A narrow

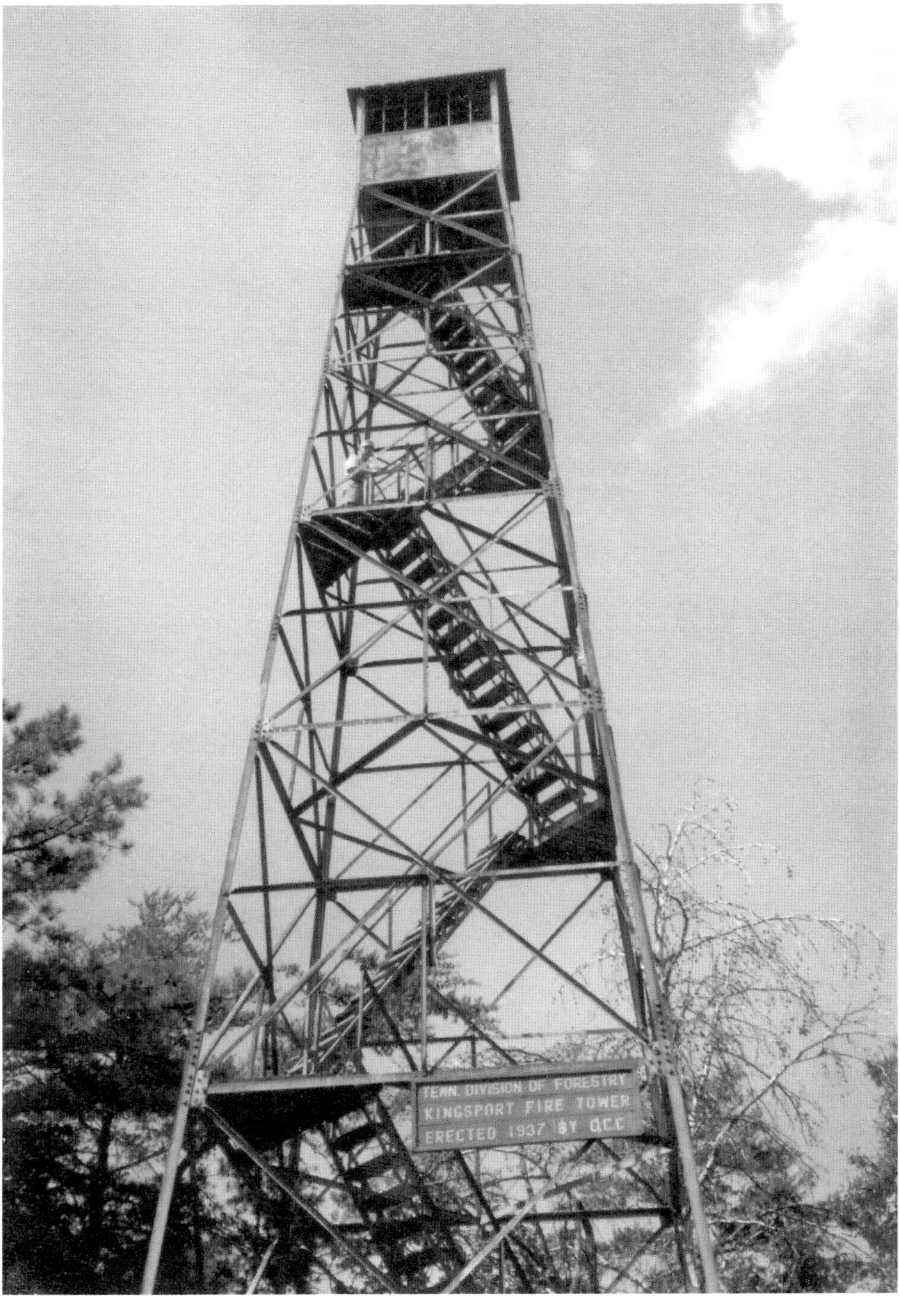

The Civilian Conservation Corps built a tower on Bays Mountain near Kingsport, Tennessee, in 1937, according to a sign on the structure. *Anne Whittemore.*

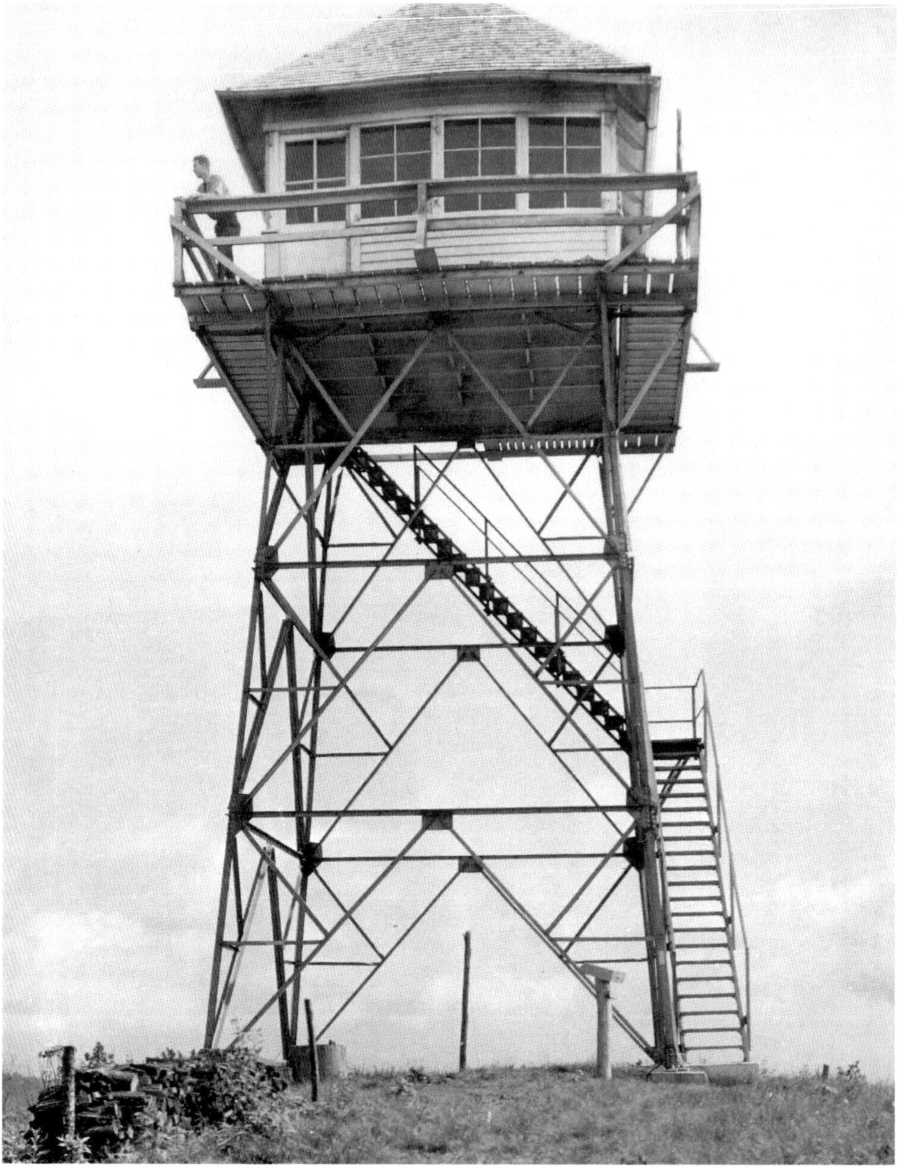

A man watches for smoke from the catwalk of the Rich Mountain Fire Tower in North Carolina. *National Archives.*

gravel and dirt road, Forest Road 467, circles the tower's base. The gravel road, which is not marked, can be reached from U.S. Highway 25 near the Appalachian Trail crossing.

Another North Carolina tower, on Albert Mountain in Macon County, is also open to the public. The tower straddles the Appalachian Trail but can also be reached by Forest Road 67, a lengthy gravel and paved route that takes travelers to several remote attractions. At an elevation of 5,163 feet,

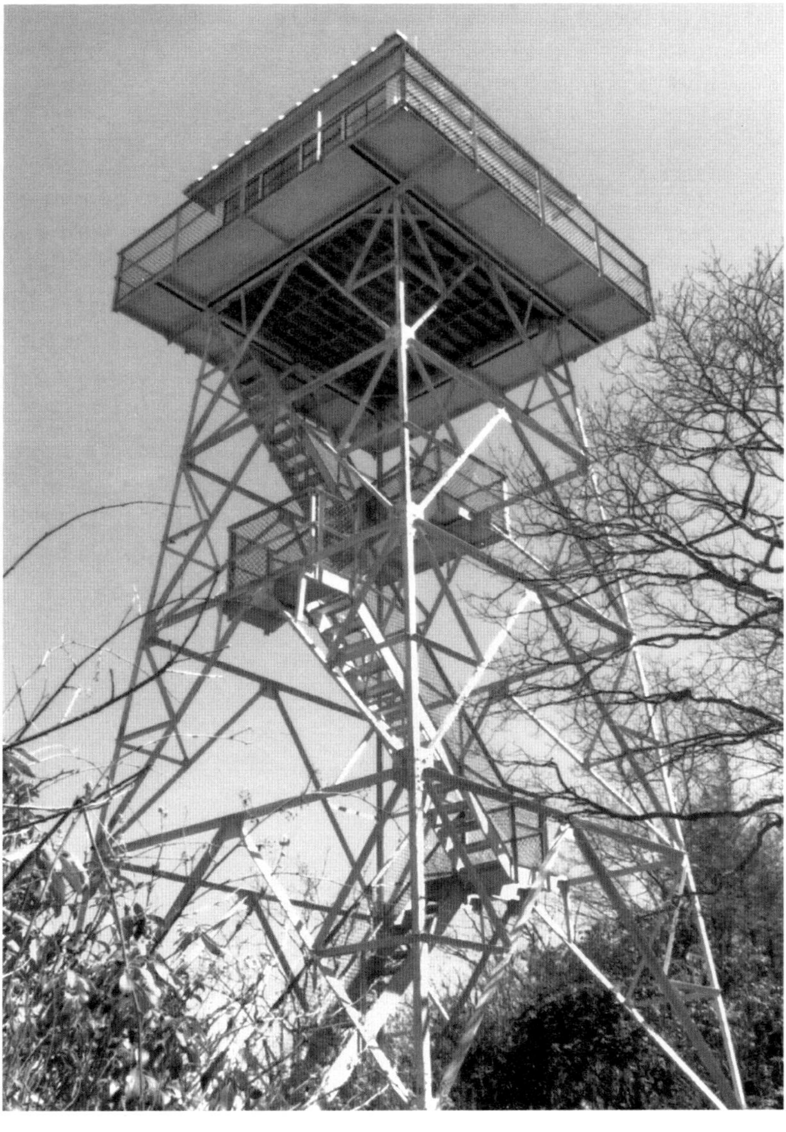

Albert Mountain Fire Tower was constructed near Franklin, North Carolina, in 1951. The forty-three-foot-tall steel tower replaced a nearby lookout at Standing Indian. *Photo by Robert Sorrell.*

according to a study by Edwards-Pitman, the 43-foot steel tower features a 14- by 14-foot wooden cab and catwalk.

The Albert Mountain Fire Tower was constructed in 1951 by the forest service to replace the Standing Indian tower. It is one of the last lookouts to be built in the Blue Ridge for fire detection. Standing Indian's tower was located northwest of the Albert Mountain tower and was dismantled in the 1950s. Standing Indian was a two-story lookout. The ground level was constructed of stone, and the second level was built of wood, like other ground cabs in the Blue Ridge.

Some towers—including the Chenocetah Tower in Habersham County, Georgia, and Panther Top Fire Tower in Cherokee County, North Carolina—are open to the public on special occasions. Chenocetah, the only remaining original stone lookout tower in Georgia (others like Brasstown Bald and Rabun Bald have been altered), welcomes visitors during Cornelia's Big Red Apple Festival, according to the Georgia Department of Economic Development.

The 54-foot-high stone Chenocetah Tower was built by the Works Progress Administration for the National Forest Service in 1936. According to the City of Cornelia, the Chenocetah Tower, on top of approximately 1,800-feet-high Chenocetah Mountain, was first dedicated on June 7, 1938, and again after World War II in memory of three forest workers who died during the war. Native granite was used to build the tower, which is named with an Indian term appropriately meaning "see all around."

This beautiful tower features a wooden observation room reached by a metal spiral staircase, and the structure is topped by an inlaid peaked slate roof. The tower was active until 1975. It was listed on the National Register of Historic Places, one of only a handful of towers in the Blue Ridge to be listed, in 1984. A few years later, the Georgia Forestry Commission reactivated the tower and began manning it during the dry season. Local residents established the Chenocetah Conservation Corps, a not-for-profit organization, in 1985 to maintain and preserve the tower and grounds.

Panther Top, which the U.S. Forest Service occasionally opens during the fall for leaf viewing, is a 30-foot-tall lookout tower about five miles southwest of the town of Murphy, North Carolina. According to the forest service, the lookout is located atop a 2,293-foot summit, making it the lowest situated tower in western North Carolina. It's also the state's westernmost tower.

The Panther Top Fire Tower, built by CCC Camp F-29 near Murphy in late 1939, uniquely features one of the only remaining original Osborne fire finders in the Blue Ridge. The tower, with a fourteen- by fourteen-foot

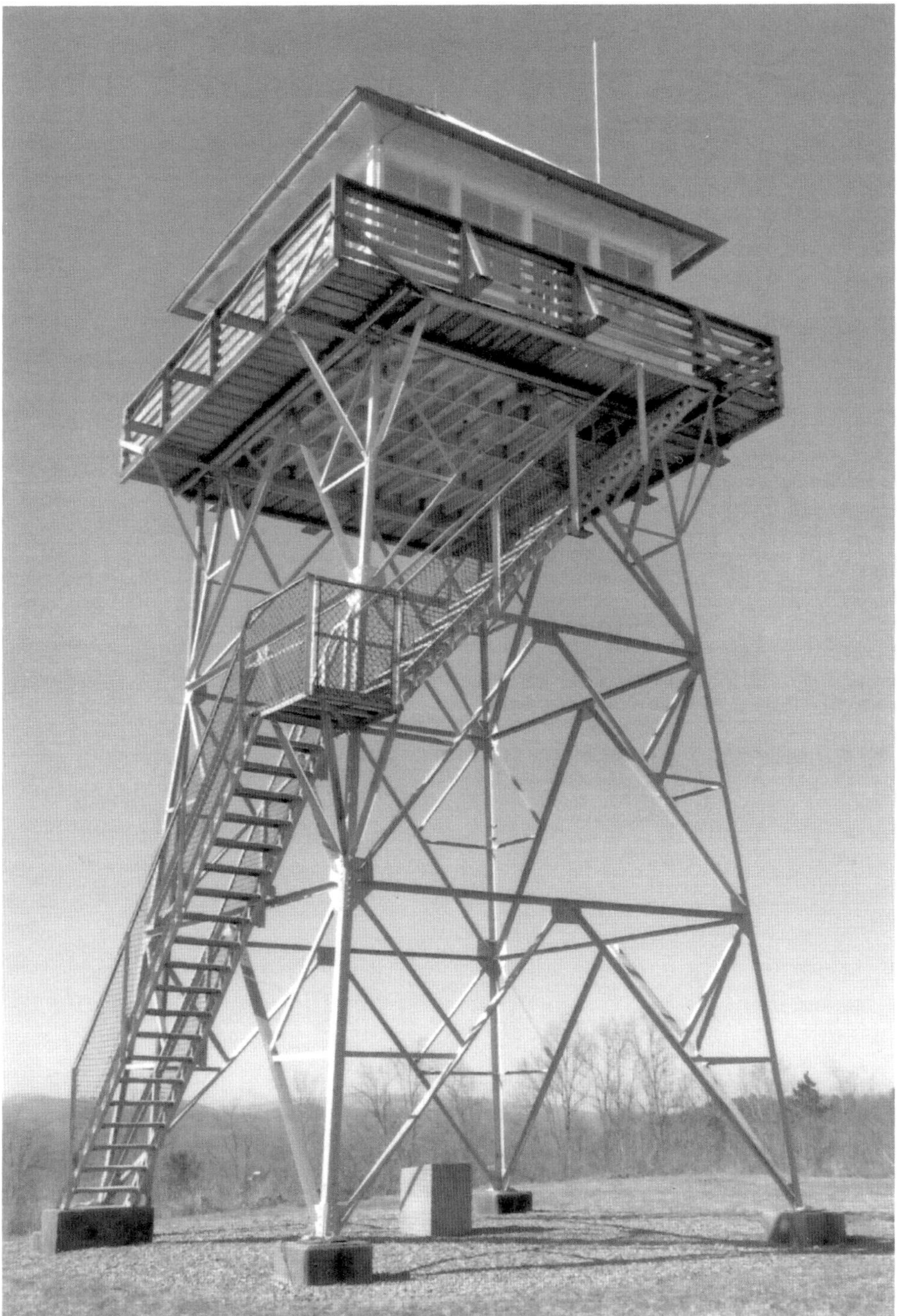

The Panther Top Fire Tower near Murphy is one of the lowest-elevation lookouts in western North Carolina. *Photo by Robert Sorrell.*

wooden cab and catwalk, according to the forest service, was last manned by a federal lookout in the 1970s. The North Carolina Forest Service has manned the tower since the 1970s.

Although Panther Top is at a relatively low elevation in western North Carolina, it has one of the best 360-degree views of any lookout in the region. Located on a grassy clearing, the tower's grounds also include "N26" markings, which could be spotted from aircraft overhead to identify the lookout.

A concrete monument, featuring a National Historic Lookout Register (NHLR) sign and U.S. Forest Service marker, is located below the tower and features the names of eleven individuals who manned the tower since 1940: Glen Beaver, Tyrone Burnette, Chester Greene, L.C. Loudermilk, Lee McLelland, Jess Rich, Ralph Rogers, Steve Rowland, Harvey Stiles, Jack Thompson and Grady Waldroop.

Edwards-Pitman, which conducted a study of several North Carolina towers in 2005, noted that Panther Top retains a high degree of historic and architectural integrity despite ongoing maintenance of the structure. "By their nature, lookout towers receive direct exposure to harsh weather conditions that exacerbate material deterioration," the study reported. "Unlike the majority of other towers included in this survey, the Panther Top tower contains most of its original furnishings."

In addition to the fire finder, the survey mentioned that the cab also contained a built-in wooden table, a metal storage box with solar batteries, a metal radio cabinet, an Artkraft insulated cabinet and a freestanding table with two chairs. And in 2002, the survey mentioned, two solar panels were installed on the south railing of the catwalk.

Unfortunately, a few of the decommissioned fire towers, even those that became public, iconic observation lookouts, have been targeted by vandals and arsonists. On Halloween 2007, the historic fire lookout tower on High Knob in Wise County, Virginia, was destroyed by arsonists, and this resulted in prison sentences for the perpetrators and the reconstruction of the iconic landmark.

"The first time I came up here on top of the knob and saw that magnificent view, I was duly impressed," said Lu Ellsworth, director of the High Knob Enhancement Corporation, which was established after the fire. He added that High Knob, a remote area above the city of Norton in Wise County, is the gathering point for residents throughout the area and has a long history. A wildfire detection facility was first built on High Knob in the 1930s, according to the National Historic Lookout Register. The original wooden

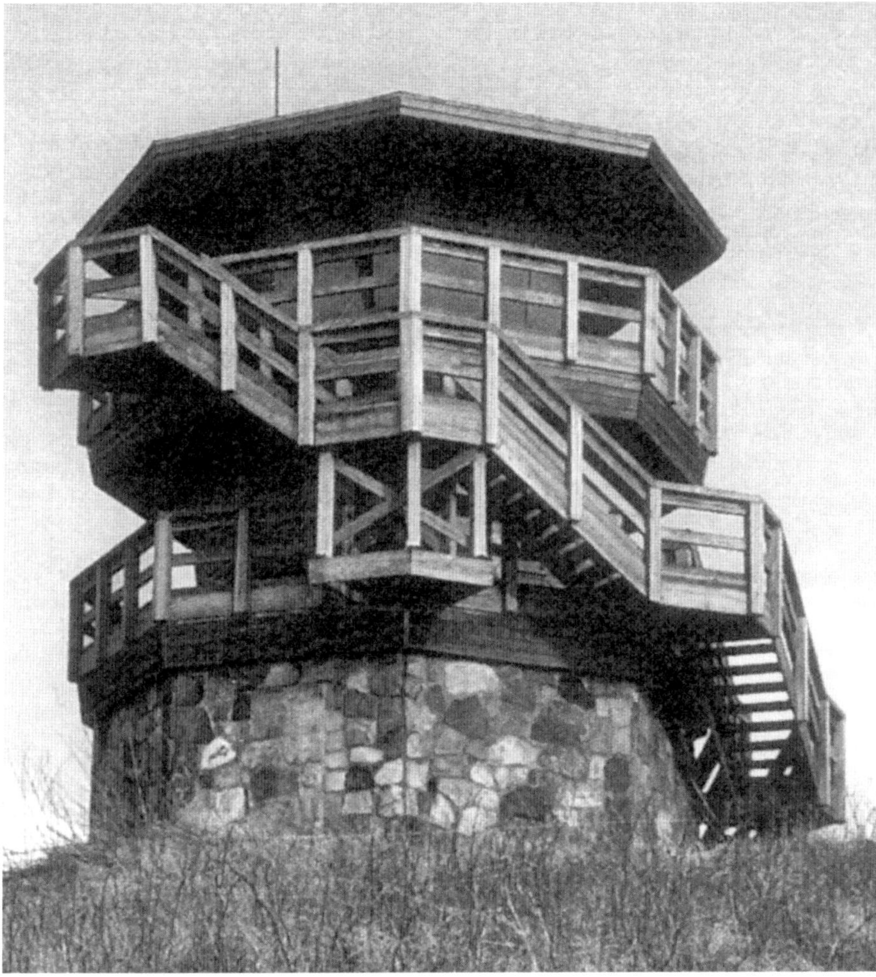

The High Knob Tower, built in the 1970s by the Job Corps, was constructed near the border of Wise and Scott Counties in Virginia. *Lu Ellsworth.*

lookout was constructed in 1938–39 by Company 2348 of the Civilian Conservation Corps, which was based in nearby Norton. The local CCC company, known as Camp Lost Creek, built a fourteen- by fourteen-foot hip-roof live-in ground house. Although the cab did not have a tower below it, the structure was still considered a lookout tower. It sat on top of the grassy bald on High Knob.

In 1978–79, the Job Corps' Flatwoods Conservation Center, which was also operated by the U.S. Forest Service, built a multi-story, eight-sided structure on High Knob. The iconic observation tower became a popular tourist attraction. The tower featured an attractive solid stone bottom with an upper story constructed of wood. During the 1990s, the forest service decked the tower, put in new steps and installed a new roof.

Throughout its history, the lookout was used for fire detection purposes. From the knob, at 4,162 feet above sea level, lookouts could see for miles. Five states could be seen on a clear day, Ellsworth said. While modern technology has been used in recent decades for fire detection, lookouts would occasionally man the High Knob tower during especially dry periods.

On Halloween 2007, the Clinch District of the George Washington and Jefferson National Forest received a call from Wise County dispatch that the iconic lookout was on fire. "I got up there around 7:30 or so," Ranger Ron Bush told the *Kingsport Times-News*. "As you know, it's a two-story structure, and it was well engulfed in flames at that time. The fire departments did a great job, but there was nothing they could really save there."

Firefighters managed to stop the fire, but it was too late. The lookout was completely destroyed, leaving only the burned stone base. The firefighters also managed to keep the fire from spreading to surrounding vegetation, which was particularly dry because of a drought.

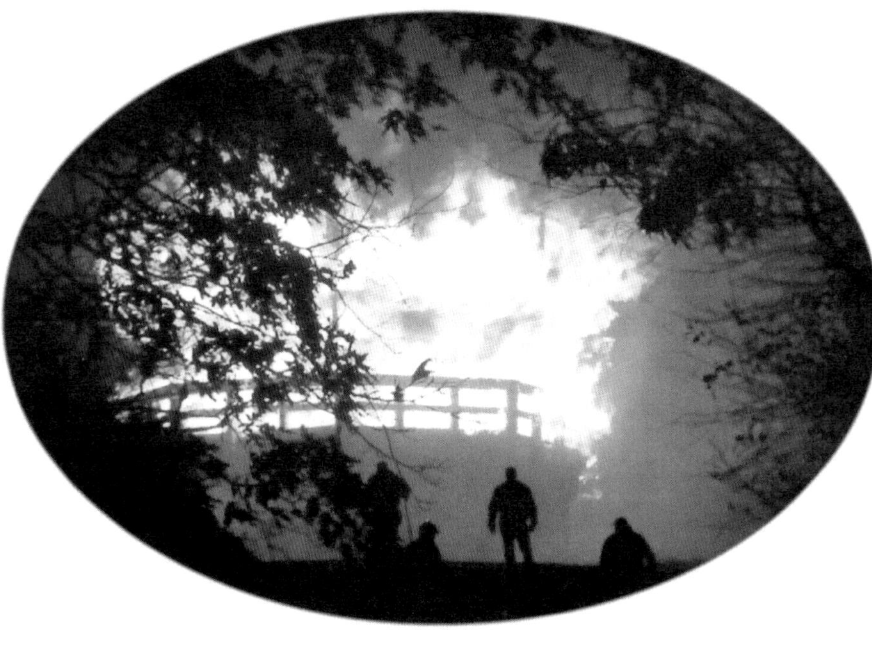

Arsonists destroyed the High Knob Fire Tower in Wise County, Virginia, on Halloween 2007. The tower has been rebuilt. *U.S. Forest Service.*

Local, state and federal authorities quickly opened an investigation into the fire, with many claiming that it was a Halloween prank. Eventually, federal authorities charged Nicholas Owens and Christopher Hyatt with arson of federal property because the lookout is located on forest service land. The two men also admitted to burning bathrooms at the nearby Hanging Rock Recreation Area. "I set fire to the High Knob Tower, I lied to investigators… and I helped set fire to the bathrooms down at Hanging Rock [Recreation Area]," Nicholas Owens said in response to a question from U.S. magistrate judge Pamela Meade Sargent in Big Stone Gap, Virginia.

According to an article in the *Bristol Herald Courier*, Owens had recently joined the Coeburn Volunteer Fire Department at the time of the fire. In federal district court, he pleaded guilty to arson of a U.S. government facility, arson on U.S. government property and making false statements to law enforcement officers. Owens was ordered to serve five years in prison and five years of probation. The maximum penalty was fifty years in prison. Owens's mother testified in court that her son, an Iraq War veteran, suffered from posttraumatic stress syndrome.

The man's co-defendant, Hyatt, pleaded guilty to arson on U.S. government property and making false statements to law enforcement. Hyatt was ordered to serve thirty-seven months in prison and three years of probation. Both men were also required to pay $523,851 in restitution for the fire.

The High Knob Enhancement Corporation was established shortly after the fire, in December 2007, with a goal to raise $500,000 to rebuild the tower, said Rita McReynolds, the organization's finance chairwoman. Construction began in 2013, and the new High Knob Tower was dedicated in August 2014. Droves of local residents and officials witnessed the dedication.

Robert Robbins Jr., vice-chairman of the enhancement board, said that there was an overwhelming response from the community to rebuild the tower after it was destroyed in 2007. "The first tower [the one-story structure built in the 1930s] lasted 40 years," Robbins said. "The tower that burned lasted for about 26 years. Let's hope this one lasts for 100 years or longer."

Rita McReynolds, who organized fundraising for the enhancement board, said that countless individuals contributed to the project, which totaled about $600,000. She recalled visiting students at Coeburn High School. "One of the students looked at me and said, 'Ma'am, I just want you to put that tower back up there where it belongs.' Well, that said it all," McReynolds said.

A wide array of people from Wise and Scott Counties got involved in the project, said enhancement board Chairman Lu Ellsworth. "On Halloween

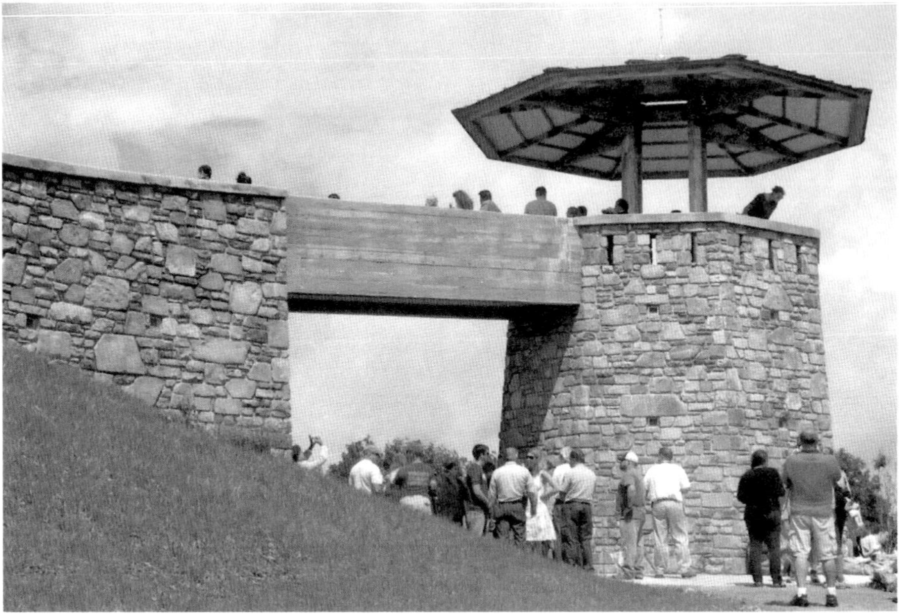

Seven years after arsonists destroyed the old High Knob lookout, the community dedicated a new structure in 2014. *Photo by Robert Sorrell.*

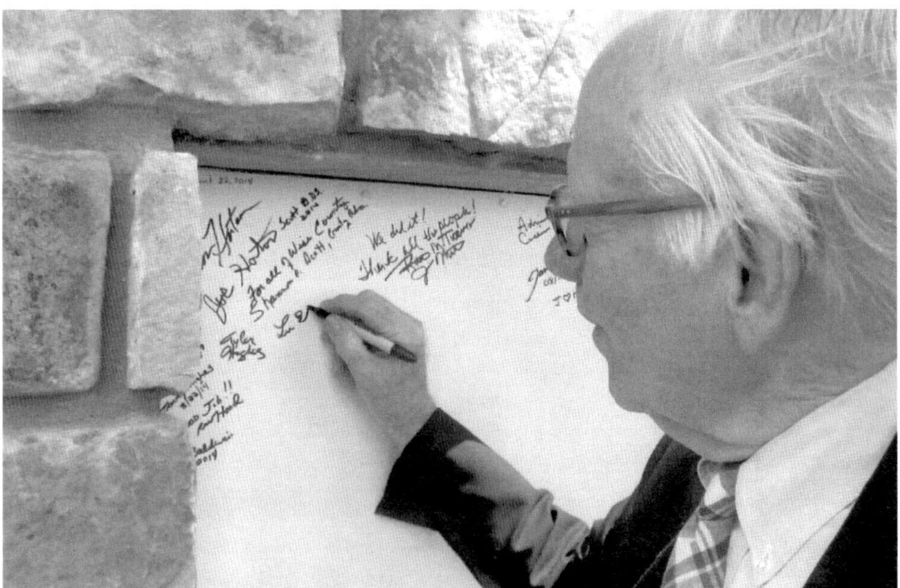

High Knob Enhancement Corporation chairman Lu Ellsworth signs his name to the wall at the new fire lookout. *Photo by Robert Sorrell.*

2007, arsonists destroyed the tower that so many people identified with and loved and had memories of," Ellsworth said. "Citizens were indeed concerned."

Ellsworth said that Robbins is the man who drove the project to success. "Right after the fire, it was Robbie Robbins who kept after folks," he said.

The enhancement board developed requirements for the new tower, Robbins said. Designed by Hill Studio P.C. of Roanoke, Virginia, the structure had to have a 360-degree view, Robbins said, so visitors could see all five states from the deck. The board also wanted the tower to be low maintenance and look as much like the old tower as possible. "I think that has been achieved," Robbins said. "We also wanted it to be a vehicle that would showcase the historical and cultural items of this region." Perhaps most important, he added, the board wanted the tower to be fireproof, so it was constructed of concrete, rock and steel.

Bill D'Angelo, of Dot Construction, which built the tower, said specific fireproof materials were selected to build the new lookout. "It won't burn," he said after its dedication in 2014, seven years after its predecessor was destroyed by arsonists.

Norton city councilman Mark Caruso, who is also an avid outdoorsman, said that the community proudly stands as a gateway to High Knob. "It is more than just an observation tower," he said. "This structure is the crown jewel of a series of natural and man-made assets in our region that are poised to change the economic outlook of Southwest Virginia." He added, "The rebuilding of this tower represents more than just a replacement of a burnt down structure. It is a symbol of hope. Like the legendary phoenix rising from the ashes, this tower has risen from the ashes as a symbol that we are rebuilding our economy, rebuilding prosperity in our region, based on adventure tourism."

Multiple arson attempts were made in the spring of 2009 at the historic Wayah Bald in Macon County, North Carolina. The local sheriff told media that his department arrested six youths for possession of weapons of mass destruction and attempted arson. The individuals were alleged to have thrown Molotov cocktails and utilized other incendiary devices at the stone lookout tower on several occasions, even videotaping one incident and posting it on the Internet. The tower survived the youths' arson attempts and even received additional federal funding for preservation and restoration.

A historic fire tower on Wesser Bald in Macon County, North Carolina, was targeted by arsonists in August 1979. Originally built by the forest service and CCC in 1936, according to an Edwards-Pitman study, the Wesser Bald Tower featured a 14- by 14-foot wooden cab. During the summer of 1979, someone decided to set fire to the live-in cab. The cab was destroyed and left only the tower's steel legs and supports. For about fifteen years, the steel structure stood atop the 4,627-foot high bald.

In 1993, the forest service constructed a wooden observation deck on top of the thirty-foot-tall steel structure, once again creating a location for viewing the surrounding landscape. The Appalachian Trail skirts Wesser Bald.

A few of the old decommissioned fire towers have been dismantled and replaced with different structures. Present-day observation lookouts at Clingmans Dome, Look Rock, Mount Mitchell and Birch Knob, in Dickenson County, Virginia, are former tower sites.

Construction of the Birch Knob Observation Tower was completed in December 2003 by the U.S. Forest Service. The tower's observation deck is at an elevation of 3,144 feet above sea level, and visitors can see Kentucky, West Virginia, Virginia, North Carolina and Tennessee. The county government noted that on a clear day, visitors can also see Ohio.

Before the new lookout was constructed, there was an old fifty-four-foot-tall fire tower on the mountain. The tower, which was in operation from 1964 to 1978, was dismantled in 1980. Piers from the tower are still visible on site.

Connie Miller of Blount County, Tennessee, recalled visiting an old steel tower near Camp Montvale in the late 1950s. Although Miller was not certain of the tower's name, Miller may have visited the old steel tower at Look Rock, which was replaced in the 1960s with a concrete public observation tower. "I spent a week at Camp Montvale in the spring of 1957 when I was in seventh grade, and during my week there, we visited the fire tower," Miller recalled.

Camp Montvale was a popular youth camp in Blount County. During the summer of 1960, Miller and her church youth group took a hike to the unmanned tower near Camp Montvale. "We arrived up top, and I was all for climbing the metal tower and entering the watch-room at the top, which was accessed through a trapdoor in the floor of the room," Miller recalled. "I knew it was not manned at the time, but I do not recall that any no trespassing signs were posted."

Miller said that she remembers climbing the stairs to reach a trapdoor, or hatch, that led into the tower's cab. "The watch-room was bare of any accessories," Miller said. "It was just a plain room, probably as small as eight by eight foot." Miller recalled it being a beautiful and hot day, and the group observed no fires from the tower.

Although she had no problem with the climb up the abandoned tower, the trip down was difficult, she said. "We joked around a bit, then opened the trapdoor and stepped onto the narrow rungs of the metal ladder, and

I froze," she recalled. "I had made the mistake of looking down, and when I realized how far below the ground was, I panicked and could not move. Our group leader and a couple of the boys talked me down the ladder, and I haven't been that high since then."

Records are far and few between for the old Look Rock steel tower, but it was dismantled as part of Mission 66 and replaced with an eighty-foot-high concrete observation tower. Mission 66, a national park service program, created dozens of modern public facilities across the country, including Look Rock and Clingmans Dome.

Through Mission 66, the National Park Service sought to modernize the system for the fiftieth anniversary of the park service in 1966. In 1955, the system's director, Conrad Pirth, proposed a ten-year plan to modernize the parks, providing better access for visitors. From 1956 to 1966, the federal government put more than $1 billion into the national parks for construction projects. During that time, Mission 66 planners and architects developed the concept of the visitor center to streamline and standardize visitor services at federal parks nationwide. About one hundred new visitor centers were built during the ten-year program.

Prior to Mission 66, most structures in the national park system used rustic architecture, which was affordable and easy to construct during the 1930s when the CCC did most of the work. Updates to the nation's parks had not been completed since the 1930s. Thomas Chalmers Vint, a landscape architect and the system's director of design and construction, called for the projects to be based on modern design.

Mike Aday, a librarian and archivist at the Great Smoky Mountains National Park, said that the modern Look Rock Tower was built in 1967 by Hart Construction Company. Look Rock's new facility was based on Clingmans Dome, which was built in the Smokies during the Mission 66 project. The park, which has limited records, does not have records that detail the original steel tower or its disposition.

Another national park official, Cultural Resource Program manager Dianne Flaugh, said that the original Look Rock Tower appeared to have been built by the CCC in the 1930s and that there was one parcel of land along the Foothills Parkway that was owned by the State of Tennessee and the state's forestry department. The fire tower may have been located on the state's parcel. Construction of the Foothills Parkway began in 1960 and continued through 1969.

Clingmans Dome, a public observation tower at Tennessee's highest point, opened in 1959, nearly one decade before Look Rock. It is one of nine such

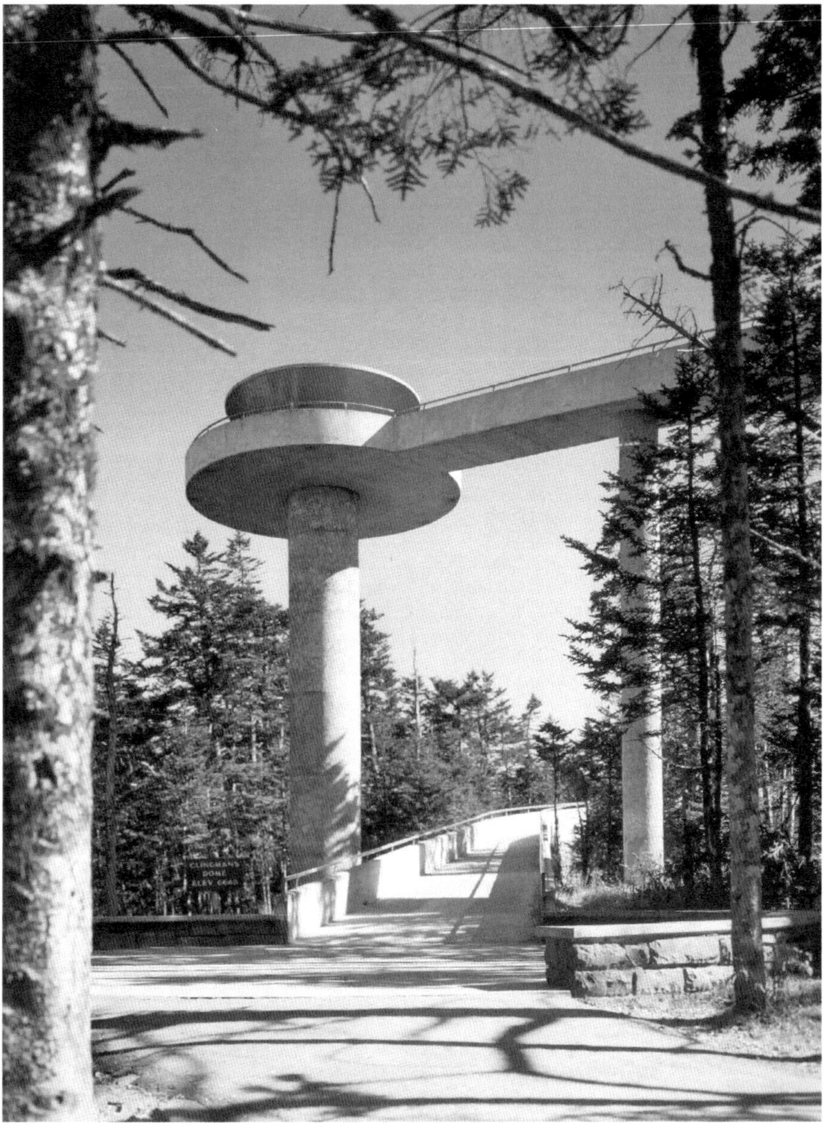

Clingmans Dome, a modern observation lookout, is one of the most popular attractions in the Great Smoky Mountains National Park and replaces a steel tower. *Library of Congress.*

observation towers built in the United States during Mission 66; a similarly designed structure was built in the Everglades National Park. The tower, with a sweeping walking ramp, was not built for fire detection but rather for the millions of visitors to the Great Smoky Mountains National Park. It had replaced a wooden tower that was built in the 1930s but removed in 1950, according to a National Register of Historic Places nomination form.

According to the form, a National Park Service press release explained that from the advantageous height of 6,688 feet (the combined height of the tower and the mountain), visitors would "have the opportunity to compare the variety of changes in mountain scenery, and to clearly observe the famed smoke-like haze." The press release went on to say that the Clingmans Dome Tower was being built as part of the Mission 66 program.

A Gatlinburg, Tennessee architectural firm, operated by Hubert Bebb and Raymond I. Olson, designed the new forty-five-foot-high observation tower. But Bebb, the primary architect, originally submitted a design that was ultimately rejected because it included a fire lookout cab. The design featured a massive central cylinder built of uncoursed masonry, or material seemingly randomly placed, and approached by a ramp of the same style. A passageway would have led from the stone ramp through the central tower and onto a concrete ramp that spirals up to the top of the stone tower. His original design also featured an enclosed fire lookout cab on top of the stone tower.

The NRHP form noted that John Cabot, the acting chief of the regional design and construction office, liked Bebb's original plans. "[T]he entire concept [of the ramp] has rather a playful, even romantic, feeling, which would be liked by a great many visitors," Cabot said in correspondence.

Fred H. Arnold, acting chief of the Branch of Park Forest and Wildlife Protection, argued against the inclusion of the fire lookout cab. According to the NRHP form, Arnold believed that although Clingmans Dome was the highest point in the park, it did not follow that it was a strategic location for fire detection. As the park system had previously developed its fire detection system, Clingmans Dome had been eliminated as a useful site. Before the 1950s, it had rarely been used for fire detection purposes.

After further discussions, the fire lookout cab was removed from the site's final designs. Officials also decided to remove the central cylinder tower from the plans, replacing it instead with a "mushroom-type cap." They determined that a heavy masonry tower was not necessary to hold a ramp that remained in the plans.

Construction of Look Rock and Clingmans Dome did not come without controversy. Both sites, designed with modern architectural styles, strayed from the park's typical rustic appeal. Even before Clingmans Dome was built, when the designs were made public, the National Parks Association, a conservation organization, fought against the new tower. The NPA called

Clingmans Dome "flashy and conspicuous," according to the NRHP nomination form.

"To the NPA, Clingmans Dome Tower seemed incongruous with the surrounding landscape and in opposition to the values of the National Park Service, while the National Park Service viewed the tower as a means to fulfill their mission by providing for visitor access," the nomination form said.

But despite the controversy, the park service received strong support from local residents and organizations. W.C. Norris of Waynesville, North Carolina, was awarded the contract and completed the tower on October 23, 1959.

The tower built at Look Rock was based on Clingmans Dome but included a fire watch cab, separate from the viewing platform. Another difference is Look Rock's switchback ramp, rather than the spiral ramp at Clingmans Dome. The setting of Look Rock's property, which is located on a narrow ridge, required a narrow access ramp.

Many decommissioned towers are not accessible to the public; either their steps have been removed or the hatches, or entrance doors, to the cabs and catwalks are locked. The tower on Holston Mountain near Elizabethton, Tennessee, is gated, and "No Trespassing" signs are posted. Despite being gated, there are some gaps in the chain-link fence that people are known to climb through to access the lookout. Wooden steps up the tower have been replaced with steel steps. The original wooden cab and catwalk on the Holston tower have been removed, leaving only a steel platform on top of the structure. Although the tower is considered off-limits by the forest service, an old road is accessible, allowing hikers to visit the lookout's base.

Doris Smithson Temple, who is no longer able to access the fire tower on Camp Creek Bald, along the Tennessee and North Carolina state line, recalled memories of visiting the landmark in her youth. "My wishes are for some of my ashes to be scattered there along the base of the tower," Temple said in 2014 as she remembered stories from the tower.

Temple recalled visiting the historic Camp Creek Bald Fire Tower with her family during the 1960s, '70s and '80s. The tower, which was constructed in 1929, once featured a fourteen-foot-square cab. In the 1960s, however, the forest service replaced the square cab with a unique round cab. The service decommissioned the tower in the 1970s.

Temple recalled riding in the bed of Millie, a red Vermillion farm truck, as her parents drove up the mountain after church for a picnic on Sundays. They would drive up Greene Mountain Road to Bald Mountain Road and then to a clearing at the base of the tower for a picnic. "After we had

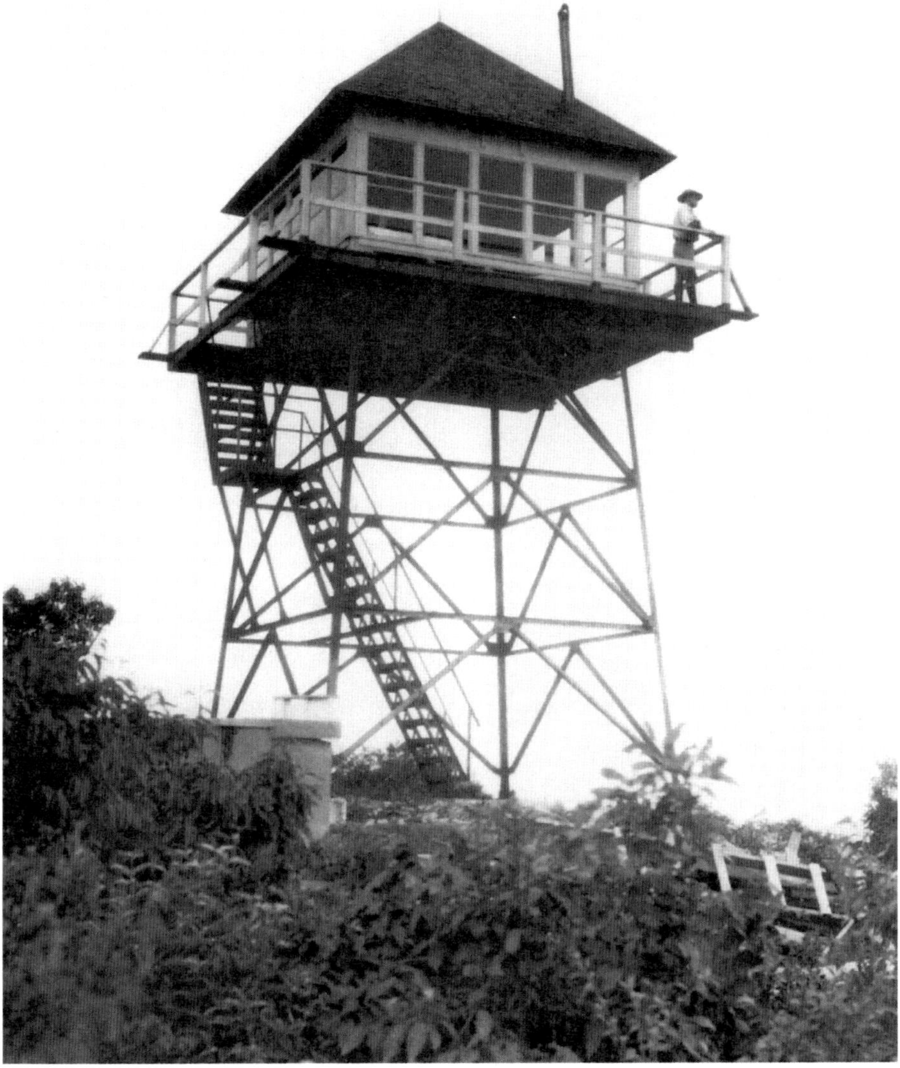

A lookout tower featuring a square cab was built on Camp Creek Bald in 1929 on the Tennessee and North Carolina border. *National Archives.*

our picnic dinner in the clearing at the base of the fire tower, we would climb the steps up to the tower and walk around the observation deck," she recalled. "While walking along the deck, Daddy would take great pride in pointing out to us all the states we were seeing from the deck."

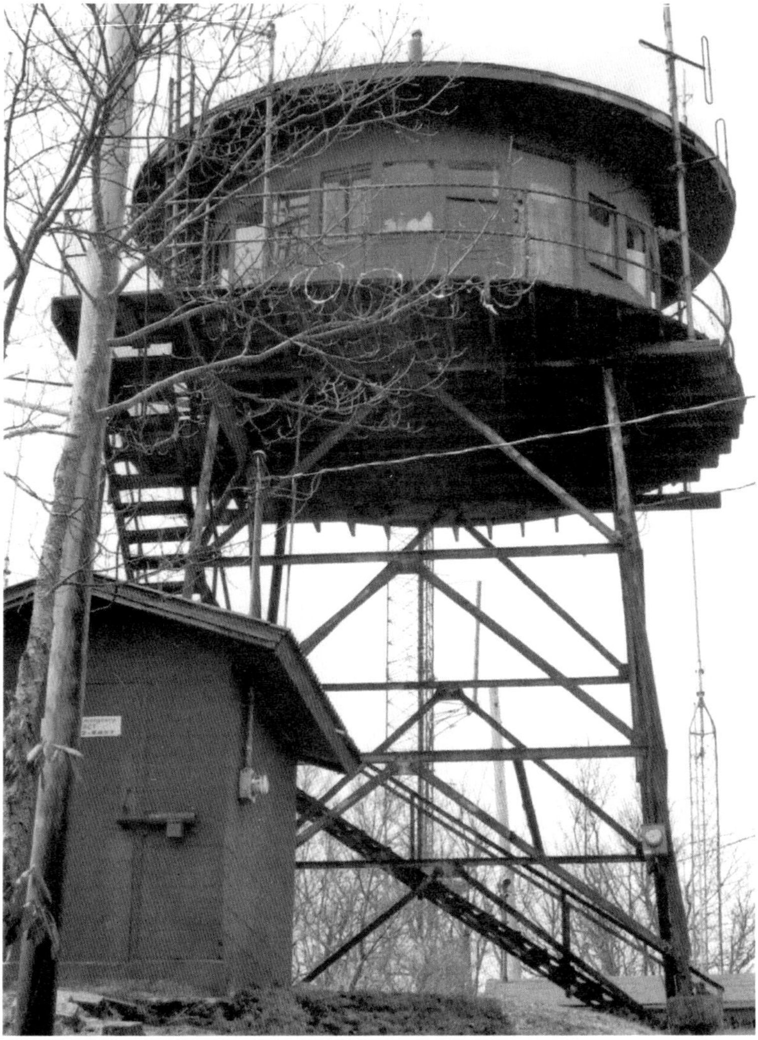

A unique circular cab is mounted on the steel supports of the Camp Creek Bald. The tower once had a square cab. *Photo by Robert Sorrell.*

They would also peer through the windows of the tower's round cab, where Bud Laws worked during fire season. "He would use the very large rotating telescope, compass, maps and binoculars in the center of the room to spot any forest fires and call them in," Temple said. "On one occasion when we were there, Bud was working. He gave us a tour of the tower and his living/working area and answered our many questions." The family would then climb down the tower and walk for a bit on the Appalachian Trail, which passed over Camp Creek Bald.

Temple remembered a story about her father making a nail to open soft drinks. The family had forgotten to pack a bottle opener, and her father hammered a nail into one of the utility poles. "My last trek up to the observation deck of the tower was in the fall of 1989," Temple said. "At that 1989 picnic, I noticed how the tower was beginning to show its age."

And then, a few years ago, Temple and her family returned to Camp Creek Bald. "I was so saddened to see how years of neglect had left the tower in such a state of disrepair," she recalled. "Seeing the wire all along the steps up to the tower to prohibit us from climbing up to the deck for yet another view of these beautiful mountains was also very sad for me."

The Ocoee Whitewater Center in the Cherokee National Forest in Polk County, Tennessee, is home to the former Oswald Dome Fire Tower. Before the lookout tower was moved to the Ocoee Whitewater Center, Delce Dyer and Quentin Bass, Cherokee National Forest staff, wrote about it in a report called *Rural Historic Landscapes and Interpretive Planning on Southern National Forests*.

"[A] tall tower such as the ninety-foot one at Oswald Dome near the Ocoee River seems dangerous for even fire spotters to climb!" they wrote. "If this structure is taken down, it is possible that the cab could be mounted

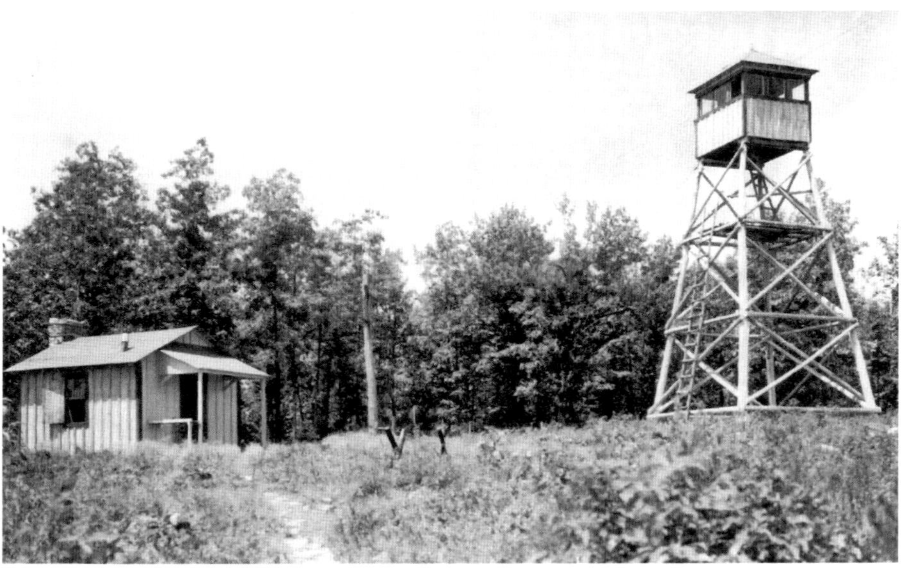

The Oswald Dome Fire Tower has been removed from the mountain and placed at the Ocoee Whitewater Center. *National Archives.*

at ground level, either outdoors or inside a visitor center, to allow forest visitors to climb through the trap door and operate the Osborne fire finder inside the diminutive space."

The tower on Oswald Dome in Polk County, at 3,500 feet, was dismantled by the U.S. Forest Service and moved the Ocoee Center. It is used as an outdoor learning facility. At its original location, there was also a wooden cabin at Oswald Dome.

Dyer and Bass also wrote about the Meadow Creek Fire Tower in Cocke County, Tennessee, which they said was in "fairly good condition, and is currently being maintained against vandalism." At the time, the structure was a large square building about 20 feet high. It featured a cinder block first level, a wooden second-level cab and a catwalk. When the lookout on Meadow Creek Mountain was first established in the 1920s, staff had to hike about two miles to reach the site. According to the U.S. Forest Service, five lookout structures have stood on the 2,875-foot mountain. "This structure could remain in place to interpret fire detection and could perhaps be retrofitted to be barrier-free," they wrote.

In 2012, Meadow Creek Fire Tower was renovated and established as a public observation tower. The project followed efforts and completion of restoration of the Pinnacle Mountain Fire Tower, another lookout in the northern Cherokee National Forest.

Occasionally, the public is provided the opportunity to rent and even purchase fire towers and lookout cabins. The U.S. Forest Service provides a listing of rental properties on its website, although most of the sites are in the Rocky Mountains, the Sierra Nevada Mountains and the Cascade Mountains. According to the service, cabins and lookouts are available on a first-come, first-serve basis and may be rented through the National Recreation Reservation Service. Permits are not issued to anyone under the age of eighteen, and those acquiring the permit must provide valid identification.

The cabins and lookout facilities, as the forest service notes, are rustic and primitive in nature. Most are in remote areas, and many do not have modern conveniences. Cabins may be equipped with bare basics—a table, chairs, a wood stove and bunks. Electricity and piped-in water are generally not available.

Parents are strongly discouraged from bringing children under twelve years of age to lookout towers. People afraid of heights or lacking physical strength should also avoid climbing lookout towers. Prior to the trip, permit holders are advised to contact the local ranger district for current conditions.

Many fire lookouts in the Blue Ridge have been transferred from the U.S. Forest Service to state agencies, organizations and private individuals, so other rentals may be available. Several fire towers in the Blue Ridge have been put on the auction block or real estate market.

In January 2014, a Lumberton, North Carolina fire tower was offered for sale on the Internet at govdeals.com, a website that sells surplus government items, including decommissioned lookouts. According to the listing, the ninety-nine-foot, nine-inch tower was used for fire spotting. "Tower is galvanized steel and was installed in 1935," the listing said. "Has an 8-by-8 foot one-room house [cab] on the top. Windows have been broken out of the house. Steps are wood and are not safe to climb. The tower is sitting on approx 9/10 of an acre of land the sale is for the tower and the land."

Most fire tower sales require the purchaser to move the tower. The Lumberton tower, however, allowed the purchaser to buy the land. According to the website, the tower and land was sold for $8,300.

In October 2014, a tower near Monticello, Georgia, just south of Atlanta, was also placed on govdeals.com. The purchaser, according to the site, required the 100- to 120-foot-high tower to be relocated. The tower was sold for $2,491. Fire towers have been sold for as little as $20, as long as the lookout has been removed from the site.

In southeast Tennessee, ham radio enthusiast John Williams built a tower in Bradley County from surplus lookouts. According to the Forest Fire Lookout Association, Williams purchased two surplus Aermotor steel towers that were located at Sparta and Sale Creek from the Tennessee Division of Forestry. Using the former towers, Williams constructed one eighty-foot-tall tower, which he uses for his ham radio operation.

For safety reasons, many of the now decommissioned towers in the Blue Ridge are being placed on lists for removal. Every state in the region is slowly dismantling these iconic structures. When fire towers become too deteriorated to restore, states and the federal government typically decide to remove them from properties. Usually deconstructed piece by piece, the material is typically used for another project, sold or discarded. The steel towers, such as Aermotor and International Derrick, are usually sold for scrap. Stone tower sites, which are usually restorable, may be removed when they become unsafe or are destroyed, such as due to the High Knob fire. The High Knob stones were incorporated into the new tower.

Once removed, the former tower sites are then usually used to construct communications towers and electric substations, or the land is left barren.

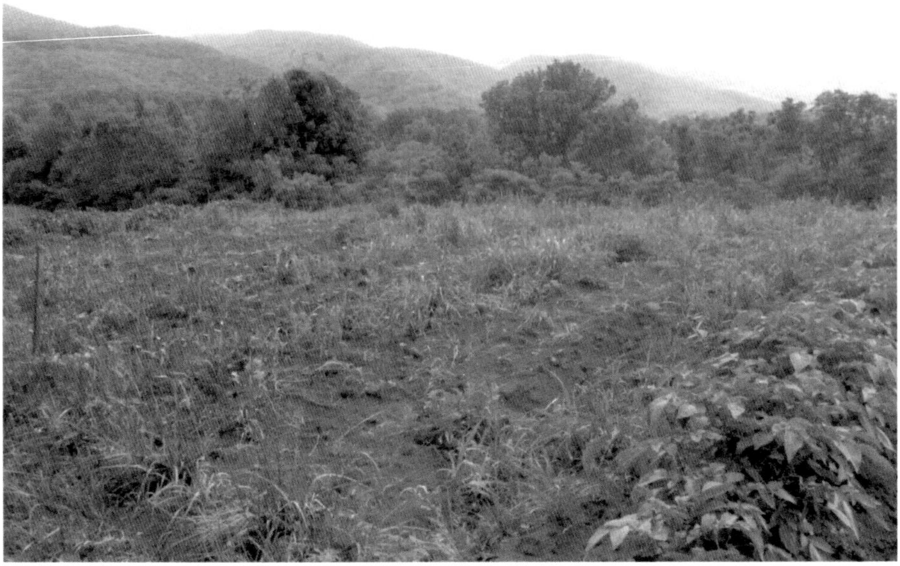

There was once a fire lookout tower on a hill outside of Elizabethton, Tennessee. The Stoney Creek Fire Tower is now a potato field. *Photo by Robert Sorrell.*

There are dozens of former tower sites in the Blue Ridge Mountains that are difficult to locate without the assistance of local residents to find the way. Old tower sites include Temple Hill in Unicoi County, Tennessee; White Rock in Carter County, Tennessee; White Knobs in Sullivan County, Tennessee; and Jasper Cliff Lookout in Lee County, Virginia, which was built in the 1950s, replacing an old CCC lookout.

An old lookout tower once stood on Buck Ridge, also known as Stone Mountain, in Johnson County, Tennessee. An old photograph of the tower reveals that the lookout featured a round cab, similar to the tower at Camp Creek Bald near Greeneville, Tennessee. Buck Ridge was manned by the U.S. Forest Service.

The former site of the Stoney Creek lookout in Carter County, Tennessee, is also difficult to locate without assistance. It was located on a hill that can only be accessed by a hollow, along Tennessee State Route 91.

Lloyd Allen, the man who purchased the tower on Little Snowball, hopes that fire lookouts throughout the Blue Ridge will be saved. Another tower, Green Knob—on which his father-in-law worked—has also been considered for removal. But forest service officials have been somewhat sentimental and have provided funding for renovations of a handful of towers, especially in western North Carolina, where the lookouts have become popular tourist attractions.

CHAPTER 5

FIRE TOWER PRESERVATION

The iconic lookout towers of the Blue Ridge Mountains fell into disrepair after state and federal governments practically abandoned them between the 1960s and 1980s. Once forest officials began monitoring the mountains from the sky, using newly available technology, the elements of Mother Nature—not to mention the hands of humans—began a slow but steady erosion. High mountaintop winds, heavy rain and snow and other weather factors have pelted the structures. Vandals have also damaged and pilfered the towers. Many abandoned lookouts are covered in graffiti or have missing steps and railings, and their sparse furnishings have been stolen or removed.

Because of vandals and dangerous conditions, government officials have installed high barbed wire fences and locked gates to many of the remaining towers in the Appalachian Mountains. "No Trespassing" and "Keep Out" signs adorn the towers in efforts to keep individuals off the deteriorating structures. Over the years, many towers have been dismantled and removed from the mountains. Many towers were removed between 1970 and 1990 after the government began using air surveillance to detect forest fires. The Great Smoky Mountains National Park, for example, removed seven steel towers in the mid-1980s but retained four for communication systems.

Still, a small handful of these iconic mountain treasures have been fully restored and are open to the admiring public as observation platforms. Through the efforts of interested individuals, organizations and communities, like the rebuilding project at High Knob Tower in southwest Virginia, the lookout towers on Mount Cammerer and Pinnacle Mountain, both in east

Tennessee, are accessible to the public and preserved for future generations. Both projects required the work of countless individuals and thousands of dollars in donations.

Citizens in the vicinity of the Great Smoky Mountains National Park became concerned about the deteriorating conditions of the stone and timber lookout tower atop Mount Cammerer in the early 1990s. Their efforts to rehabilitate the scenic tower have become a model for other restoration

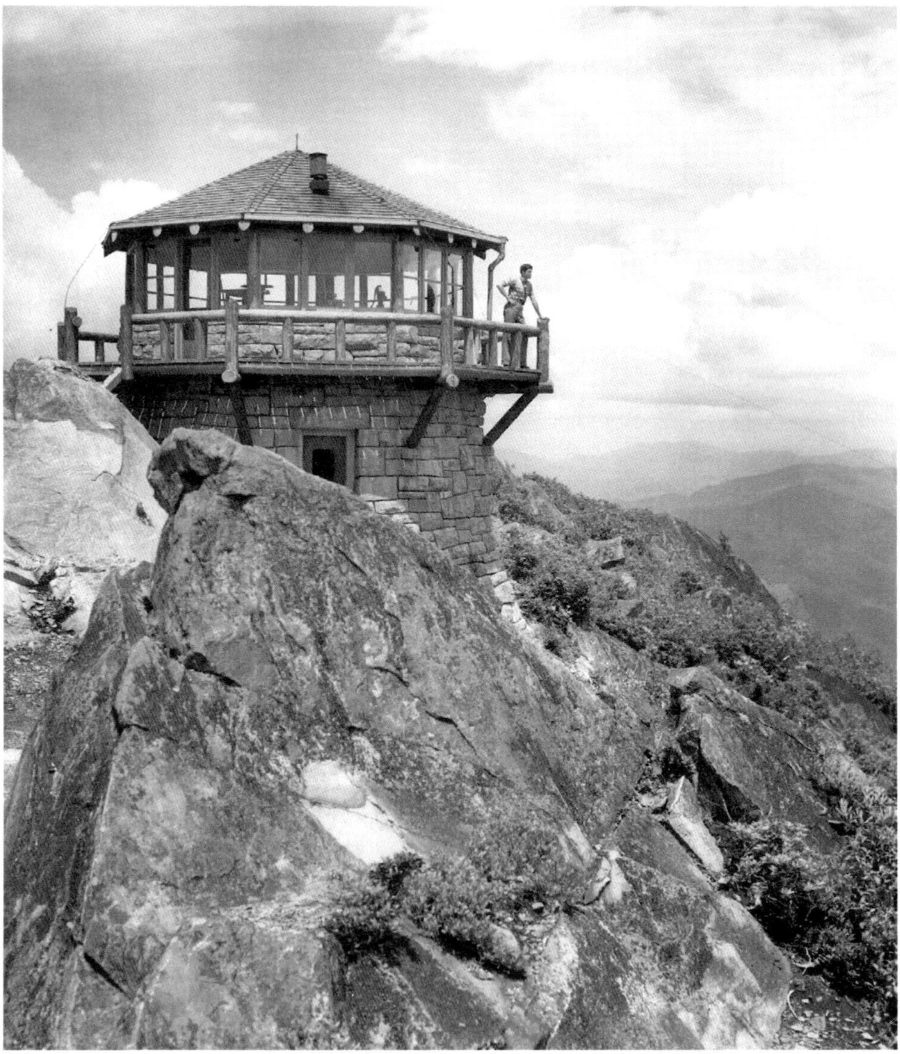

A lookout watches for smoke from the tower on Mount Cammerer in the Great Smoky Mountains National Park. *Tennessee State Library and Archives.*

projects across the country. Following its rehabilitation in the summer of 1995, Mount Cammerer became one of the most popular lookout attractions in the United States. It is one of the most visited lookout towers in the Great Smoky Mountains behind the Clingmans Dome Observation Tower. The modern tower on Clingmans Dome was constructed in 1959 but was used exclusively for park visitors and not as a fire lookout point.

Mount Cammerer's restoration saga began around 1990, according to Tennessee Supreme Court justice Gary Wade. The saga began after a few friends, Thomas Trotter and Wade, challenged themselves to a strenuous twelve-mile hike in Cocke County, Tennessee. "My friend called me and said we had reached forty years of age and had not hiked all the trails in the Smokies," Wade recalled.

In 1990, Trotter, a prominent Gatlinburg architect and Wade's close friend, suggested that the two men set aside a few days each month to hike every trail in the Smokies. Wade, a physically conscientious individual, agreed to the challenge. "He [Trotter] acquired a map of the trails and over the next two or three years charted each of the hikes we made," Wade said.

The two men began by hiking the more popular trails in the Smokies, including the Chimneys, the various routes to Mount LeConte and Charlie's Bunion. "Tom loved the Elkmont area and made sure I was exposed to every hiking trail from Sugarland Center to Tremont, such as Jake's Creek, Huskey Gap and Blanket Mountain," Wade said. "I, on the other hand, had spent much of my childhood near the Big Greenbrier entrance to the park east of Gatlinburg and was eager for Tom to experience the Ramsey Cascades, Maddron Bald and Porter Flats trails."

One hike, Wade said, stood above all others: the Low Gap trail from the Cosby Campground to the top of Mount Cammerer, a place Trotter had never seen. "Despite the degree of difficulty, I wanted Tom to see what my brother and I had experienced a few weeks earlier," Wade said.

Wade and his brother, Dr. Dwight Wade, had been to the lookout tower on Mount Cammerer prior to Trotter's visit. The hike, which begins at the Cosby Campground in Cocke County, follows the three-mile-long Low Gap trail. Wade said that the hike, especially the Low Gap trail, is a fairly difficult, steep hike. Tower hikers then follow the Appalachian Trail before reaching an approximately half-mile spur trail across Mount Cammerer's rocky ridge to the tower.

The Civilian Conservation Corp built the octagonal, mountaintop, western-style tower in the late 1930s after the Great Depression. It provides visitors with 360-degree, unobstructed views of the Smokies. The tower

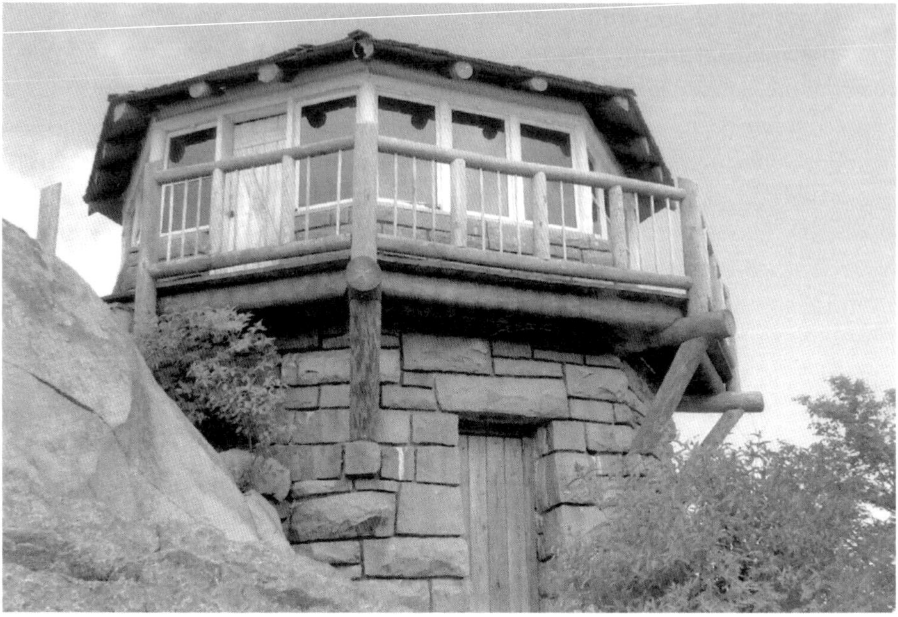

The Friends of Great Smoky Mountains National Park organized efforts to restore the lookout tower on Mount Cammerer (seen here) in the 1990s. *Robert Sorrell.*

features a stone foundation. Some of the stones were huge and weighed about six hundred pounds, Wade said. The CCC had taken the large stones from a mountainside quarry about one hundred feet away from the tower site. The unique two-story tower, with a fourteen-foot-diameter live-in cab, was decommissioned by the national park service in the 1960s. Oddly, though, the park service renovated the tower just four years before it was decommissioned. During the 1960s restoration project, it received a new floor, a roof and other fixtures.

"Tom immediately recognized the architectural significance of the tower," Wade said of Trotter's first visit to Mount Cammerer. "He was delighted to see the western-style tower, but he was appalled by its condition."

During the 1980s, the park service demolished several other towers due to their deteriorating conditions and lack of need, but it left four standing: Mount Cammerer, Shuckstack, Cove Mountain and Mount Sterling. According to files kept by the park, historian Edward Trout informed the park's chief of the Resource Management Division in a 1985 letter that the tower lacked any historic value. Despite Trout's belief that the tower on Mount Cammerer was not unique at the time, he believed that it was the only historic tower left in the park. Trout mentioned the CCC's national significance as a reason to retain the tower.

By 1987, the park service had decided to retain the structure for administrative and operational purposes but lacked funding for maintenance and restoration.

By the time Wade and Trotter visited in the early 1990s, the walls were adorned by graffiti, and the physical structure had deteriorated. Weather conditions over the years had also damaged the mountaintop tower. The tower was built on a rocky outcropping, and this had resulted in the exposure to extremely high winds. The once spectacular structure, Wade said, was in need of restoration.

Still, the two friends enjoyed a small picnic on the adjacent rocks and by chance were joined by an unidentified backcountry ranger who had been hiking along the trail. This ranger, in talking about the tower's poor condition, mentioned that the park's superintendent, Randy Pope, was interested in having something done to the structure. Funds to repair the dilapidated tower were scarce. Trotter handed the ranger his business card and asked him to pass it along to the superintendent. "If Randy Pope really wants to do this, tell him that I will donate my time for the architectural drawings and ask him to give me a call," Trotter told the ranger.

A few days later, Superintendent Pope called Trotter and arranged a meeting. During that meeting, Pope explained that the U.S. Congress did not appropriate sufficient funds for capital improvements, including the costs associated with maintaining and repairing historic structures, such as the fire tower. The Department of the Interior and National Park Service are frequent victims of federal budget cuts.

Trotter, a successful architect based in Sevier County, Tennessee, donated his designing talents to the project. Trotter also informed the park superintendent that his friend Wade would help raise the $35,000 for construction costs. Wade, however, was a former Sevierville mayor and had just been appointed a judge for the Tennessee Court of Criminal Appeals. Judicial officials generally refrain from raising funds. As a result of Wade's dilemma, the superintendent suggested the establishment of a new organization to raise funds for the restoration of Mount Cammerer's tower.

On September 3, 1993, Superintendent Pope and park official Bob Miller invited a collection of people from Tennessee and North Carolina to form the founding board of the Friends of the Great Smoky Mountains National Park. Prominent Knoxville philanthropist Lindsay Young; Western Carolina University chancellor Barney Coulter; Stephen Woody of historic

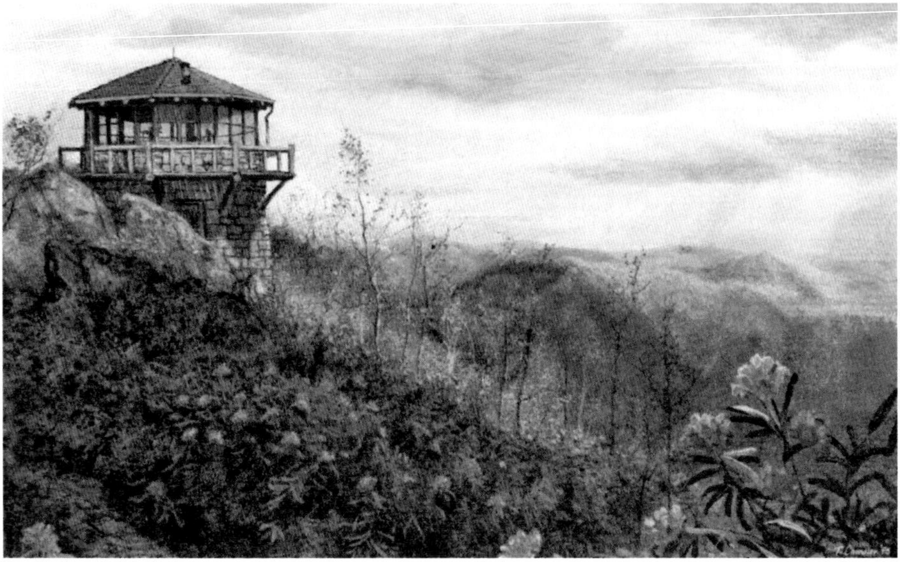

Gatlinburg, Tennessee artist Terry Chandler created a painting of the Mount Cammerer Fire Tower to raise funds for the lookout's restoration. *Gary Wade.*

Cataloochee, North Carolina; Judy Morton of Morton's Overlook; Kathryn McNeil, who made a substantial donation of land to the park in 2000; environmentalist and real estate broker Mary Johnson; and Natalie Haslam, whose family owns the Pilot Corporation, attended the kickoff meeting. Several other affluent individuals also participated in future meetings, including banker and outdoor photographer John Dickson, legendary Appalachian author and historian Wilma Dykeman Stokely, former Tennessee Valley Authority chairman John Waters, University of Tennessee vice-president Jack Williams, businessman David White and Cherokee magistrate Leon Jones.

The Rotary Club of West Knoxville stepped forward and assisted with the tower restoration project. Bob Ely, a Rotary Club officer, persuaded Gatlinburg artist Terry Chandler to donate an original painting of the tower to raise funds, Wade said. Chandler also donated two thousand prints. "The sale of Terry's prints largely covered the cost of the tower's restoration," Wade said. "We were able to raise enough funds to restore the fire tower."

The prints and paintings brought in about $25,000 for the project. Lowe's Home Improvement in Asheville donated a truckload of lumber and other supplies for construction.

In March 1995, crews airlifted two five-thousand-pound loads of construction materials to the tower site on Mount Cammerer, which is more than ten miles from the nearest road. According to Wade, Stokely Enterprises' William Stokely III, of the Mount LeConte Lodge, provided a helicopter for the project. The national park's Historic Preservation Crew began work on the tower in May 1995. "They have a very capable, skilled crew to work on historic preservation projects," Wade said.

Trotter described some of the issues pertaining to the tower's restoration: "I do recall that in preparing the drawings, historic correctness versus life safety was a big consideration. The major issue being the guardrail that wraps the tower at the catwalk or observation deck."

Trotter was concerned that the original railing height and the picket spacing did not conform to existing building code standards. "While my details weren't followed exactly, the finished product was a big improvement over the original," Trotter said. The architect also complimented the crew that restored the tower. "I was very pleased with the restoration work," he said. "It had been carried out by park staff, which is and was very qualified for that type of historic work."

The crew's four masons and carpenters camped on the mountain in order to complete the work on the tower, which included replacing windows and adding a new roof, new flooring, a reconstructed catwalk and a staircase. The restoration was completed later that summer, and a July dedication ceremony was held to honor its completion.

Several members of the Friends organization, government officials and citizens hiked up the mountain to participate in the ceremony. In 2008, during a fifteen-year celebration hike to the tower, Wade, Trotter and about thirty other individuals discovered that its restoration was needed again. "We realized that the trail had gotten a lot longer, and the grade had gotten a lot steeper," Wade joked about the long hike.

Since Mount Cammerer's 1995 restoration, the Friends group has raised more than $44 million for the park. Funds have been used for a variety of projects. Wade said that funds will likely have to be used for the iconic fire tower about every twenty years in order to preserve the structure.

Because the wood tower is exposed to the elements, it will always require some sort of restoration or replacement. Trotter and Wade made a contribution to the Friends group in 2013 to accommodate some of Mount Cammerer's ongoing work. The Friends group has also considered rehabilitation at the park's other towers at Shuckstack, Mount Sterling and Cove Mountain. "The other three remaining towers are cultural resources but do not have any special designations bringing them to the top of our restoration priorities at this time," said Dana Soehn, public affairs officer for the Great Smoky Mountains National Park.

Soehn noted, however, that "Mount Cammerer fire tower is a particularly unique stone structure that seems almost out of place in the mountains. Typically, this type of structure was built near the coast, but not in the mountains. It is not only a significant resource to the park, but also region wide and was a high priority for restoration."

Peter Barr, director of the North Carolina chapter of the Forest Fire Lookout Association, has advocated for the restoration of historic Shuckstack Tower in Swain County, North Carolina. The fairly isolated tower is located within the national park but has not been recognized for funding, as noted by Soehn.

Barr has launched the Save Shuckstack campaign to restore the more than 60-foot-tall steel tower. According to Barr, the Public Works Administration constructed the lookout tower in 1934 on the 4,020-foot-high mountain. It was manned by a lookout for nearly thirty years. Shuckstack, named because of its visual similarity with cornstalks, was built about ten years before the construction of the nearby Fontana Dam, which could be seen from the tower. The lookout, or watchman, could see the flooding of the Little Tennessee River and development of the new Tennessee Valley Authority reservoir from the tower's cab. Barr said that the tower, located a short distance off the Appalachian Trail, has sustained vandalism and weather damage since it was abandoned in the 1960s.

The park service had constructed a one-room log cabin on the ground at Shuckstack Fire Tower. The lookout would live in the cabin when not on duty in the tower. The park eventually removed the cabin, but the tower was left standing because it had been used as a radio relay point. A chimney and a cistern are all that remains of the cabin.

The park service, Soehn said, discourages individuals from climbing Shuckstack. Even Barr has noted its deteriorating, unsafe conditions. The wooden planks have rotten and cracked, Barr wrote on his Save Shuckstack website. "A hand railing on its initial flight of stairs has fallen off or been stolen, creating a frightening and dangerous climb to the top of the tower," he said. "I was so horrified by the condition of the tower that it took me nearly a half hour to muster the courage to climb it."

Upon Barr's first visit to the tower, according to his website report, his hiking partner refrained from climbing Shuckstack and decided to stay on the ground. The cab has considerable damage, such as gaping holes and broken windows. However, Barr was impressed by Shuckstack's beautiful views of the surrounding mountains and valleys.

Barr has written a book, conducted speeches and talked to officials with the national park service about restoring Shuckstack. If the tower is not restored, he said, it will likely receive the same fate as seven other national park towers: dismantlement. The park has considered placing the remaining towers on the National Register of Historic Places, and this could result in their complete restoration. Barr's efforts have raised several thousand dollars, but not enough to restore the tower.

Another Swain County tower was dismantled in 1985, but its lookout cabin remains on site. As of March 2014, the High Rocks cabin is in need of restoration, according to several sources. The cabin is located on a 5,185-foot-high mountain in the Great Smoky Mountains National Park. The CCC constructed the tower and the cabin between 1935 and 1936.

Laura Ingle, a Clemson University historic preservation graduate, wrote in 2011 about the towers of the Smokies for her thesis, "Every Day Is Fire Day: A Study of Historic Fire Towers and Lookout Life in the Great Smoky Mountains National Park." She described both the construction of the tower and cabin, as well as the issues surrounding the restoration of the High Rocks cabin, with excerpts from park letters and e-mails.

An October 9, 1935 letter, she cites, reported that seventeen men worked on the tower and cabin. They set up tents to provide shelter while constructing the site. The High Rocks Fire Tower was forty-six feet tall, according to Ingle, which was lower than other park fire towers. The national park and TVA removed the High Rocks Tower in the 1980s. They decided to retain the log cabin, making it the only fire lookout cabin remaining in the national park.

In 2004, a donor offered to pay for the costs to restore the historic cabin. The park, however, rejected the offer. E-mails, detailed in Ingle's thesis, noted that the park was not interested in the donor's assistance and did not have plans to restore the cabin. The donor had even offered to do the work himself, but the park service denied his request. The donor then offered to protect the cabin with a tarp. Park staff admitted, Ingle reported, that the cabin is eligible for the National Register and deserves preservation, "but that funding and personnel limitations prevented them from restorations."

Ingle's thesis mentions that one park service official stated in an e-mail that he hoped the donor would consider helping with another project, rather than the High Rocks cabin.

The cabin has received no restoration since the tower was removed in the 1980s. "Though it was a historical site, the cabin was in awful shape with trees growing through the floors," wrote a hiking blogger, Nick Thompson of Backpacker's Resource, in late 2013. "The best way to describe it was that it was in shambles. The roof had caved in and the only thing somewhat standing was the porch."

A fire tower in the Cherokee National Forest in Tennessee was restored in 2011. During the 2000s, residents and officials in Unicoi and Washington Counties began efforts to restore the Pinnacle Mountain Fire Tower, which is located atop 3,420-foot-tall Buffalo Mountain in the Cherokee National Forest. Subsequent restoration efforts mirrored Mount Cammerer's rehabilitation a decade earlier. The 40-foot-tall tower was constructed by the Civilian Conservation Corps in the 1930s.

From the top of the tower, which featured a living quarters, a lookout could view the forests of Unicoi and Washington Counties. The lookout stayed atop the towers for hours. The living quarters featured a cot, a refrigerator and a cooking stove, in addition to fire watching equipment. In 1949, as forest service records indicate, it was equipped with a telephone. In 1989, however, the U.S. Forest Service abandoned the tower and instead began using aerial observation methods to watch for fires. For more than a decade, the tower deteriorated due to the elements.

In 2007, the Partners of the Cherokee National Forest launched a campaign to raise a portion of the estimated $100,000 needed to rehabilitate the tower. The original plans were to replace the enclosed fourteen- by fourteen-foot cab on top of the fire tower with an open observation platform. According to an article published in the *Johnson City Press*, officials hoped to have the tower open to the public in 2008 or 2009.

Ultimately, the forest service and local tourism officials envisioned the Pinnacle Mountain Fire Tower as the centerpiece of a proposed hiking and mountain biking trail system on Buffalo Mountain linking Erwin to Johnson City. The hiking and biking trail was eventually created between the town of Unicoi and the tower. The trailhead is located conveniently adjacent to exit 32 along Interstate 26.

Unicoi County mayor Greg Lynch told a *Press* reporter, "Half our land belongs to the Forest Service. We used to see that as a liability. Now, with the growing interest in heritage tourism and recreation, we see that as an asset."

The Partners of Cherokee National Forest modeled its efforts after the Friends of the Great Smoky Mountains National Park and its successful restoration of Mount Cammerer in 1995. During a three-year period, the Partners worked with the town of Unicoi, other organizations and private citizens to raise funds for the tower. One fundraising event

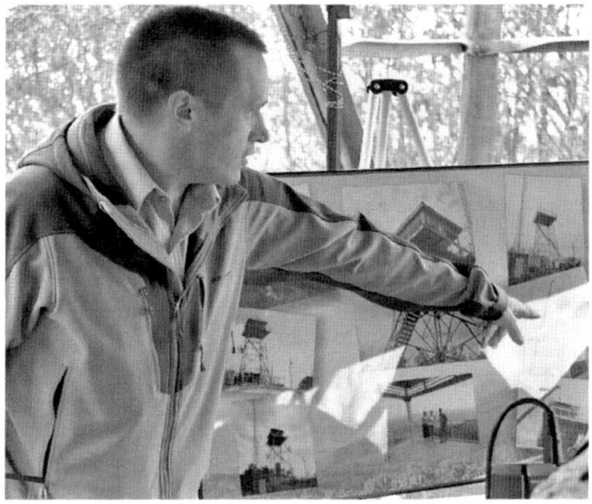

A U.S. Forest Service official views photographs of renovation work at the Pinnacle Mountain Fire Tower in the Cherokee National Forest. *The Erwin Record.*

was held in November 2009. The Partners, in cooperation with the Tanasi Arts and Heritage Center and Farmhouse Gallery, hosted a "Tower Raising" event at the Farmhouse Gallery in Unicoi. Citizens also contributed thousands of dollars to the Pinnacle Fire Tower Fund through Mountain Commerce Bank.

"This is a tremendous project with community leaders, individuals, businesses, groups and organizations in the area coming together for a common goal," said former Cherokee National Forest supervisor Tom Speaks in 2010 as the tower's restoration was moving forward. "I appreciate the dedication and hard work of all those who have been involved with this project. The Partners of the Cherokee National Forest and various groups and community leaders in the area have played a very important role of helping to lead the way and for building support for the project. This is the type of project that we can all be proud of."

A U.S. Forest Service press release, issued in April 2010, reported that the first phase of the project was to evaluate the existing condition of the tower and develop the architectural design for its rehabilitation. In 2010, those plans were finalized so that the reconstruction phase could begin later in the year.

Greg Lynch, Unicoi County's mayor, noted:

> *Pinnacle Lookout Tower has become a well known icon in Unicoi County and the local region. I am very interested in this project and recognize both the historic value and potential economic impacts of this project. The partnerships that have been established as a result of this effort are amazing. People have not only contributed funds but time and work on the ground to this effort. I consider this an outstanding example of a community pride project. This is going to be something that people in Unicoi County and northeast Tennessee for that matter will utilize and be proud of in the years to come.*

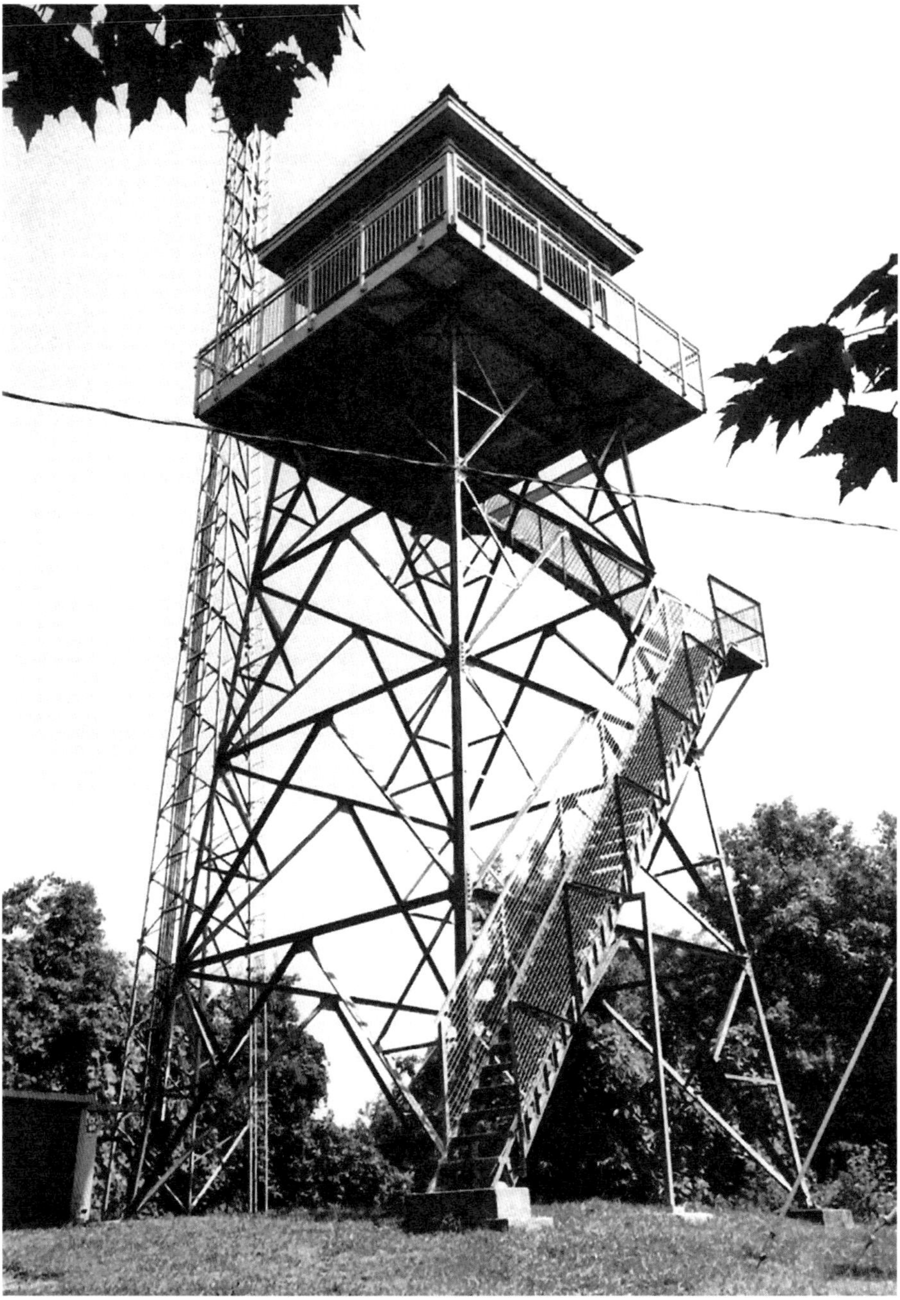

The Pinnacle Mountain Fire Tower has been completely restored and features a metal, grated observation deck. *Photo by Robert Sorrell.*

The restoration project was completed in September 2011, when officials reopened the tower to the public. During a special event, officials and citizens climbed the forty-foot-tall tower and stood on its steel grating observation deck built to replace the original fourteen- by fourteen-foot enclosed cabin. A crane had been used to remove the cabin. From the observation deck, visitors can see the town of Unicoi, Unaka Mountain, the Cherokee National Forest and portions of Erwin and Johnson City.

"If this tower could talk, imagine the stories it would tell," said Town of Unicoi mayor Johnny Lynch. "All the folks at the foot of the mountain are elated. A lot of people wanted this done."

Erwin resident Fred Hardin manned the Pinnacle Fire Tower for about eight years during the 1980s, the last decade the forest service used the structure to observe fires. "Whenever they needed you, you had to go up there and you had to stay the night," Hardin told *Press* reporter Brad Hicks in January 2014. "As long as they had a fire, you had to stay up there."

According to forest service records, the tower was refurbished in 1984. Five years later, however, the agency discontinued manning the tower. The Pinnacle Tower was one of the last towers manned in the Cherokee National

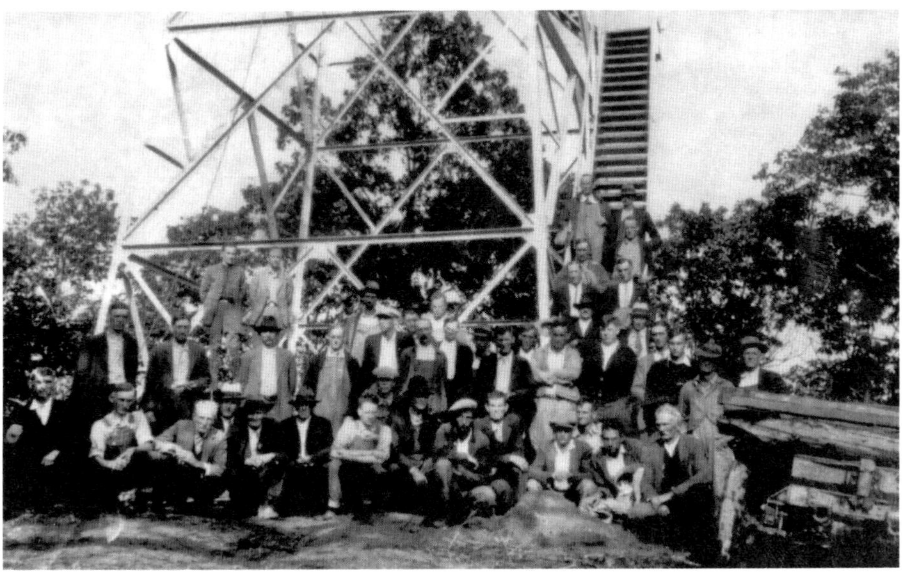

A group of men poses in front of the Civilian Conservation Corps–constructed Pinnacle Mountain Fire Tower in Unicoi County, Tennessee. *Beth Bradford.*

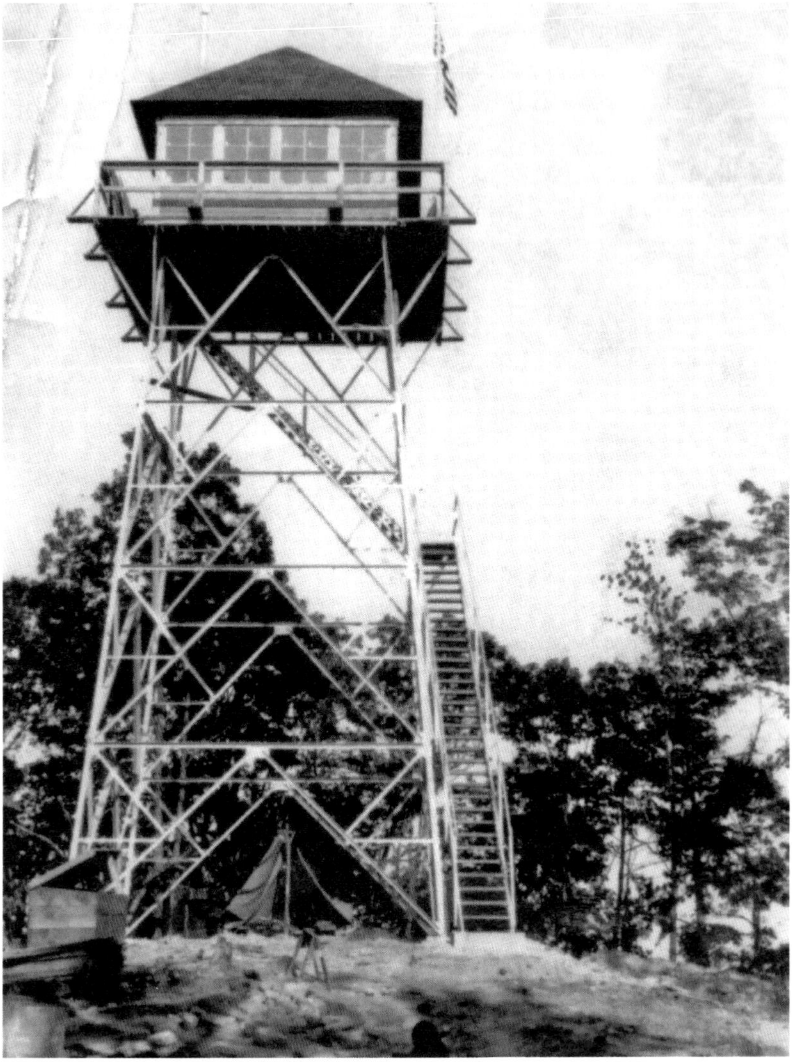

The forty-foot-tall Pinnacle Mountain Fire Tower was built in the 1930s. It is located above the town of Unicoi, Tennessee. *Beth Bradford.*

Forest. Many towers had already been abandoned in the 1960s and 1970s. After Pinnacle Tower was abandoned, vandals gained access to the tower and damaged the cab and the wooden structure.

The Partners of the Cherokee National Forest and local citizens eventually sought funds to restore the structure and converted the tower into an observation platform. Gary Schneider, president of the Partners group, said that the Pinnacle fire tower is the most ambitious project the organization had completed at the time.

The nearly 4.8-mile hiking trail leading up the mountain to the tower was created and flagged in early 2008. In the middle of July 2008, volunteers Ed Oliver and Jim Foster tied their last pink ribbon to the last tree branch. The five-month-long flagging was finally complete, according to an article published in the *Erwin Record*. "While they worked slowly, sometimes crawling on their hands and knees or stepping over 'blow-downs,' fallen trees, they saw wildflowers bloom and aged trees burst into an emerald green," noted the article.

The four-foot-wide dirt trail has several switchbacks, but it is a fairly easy hike due to the low grade. There are several points on the trail where hikers can view nearby mountains, Interstate 26 and the town of Unicoi. The tower, at the end of the trail, offers 360-degree views.

A parking area and public restrooms have been built at the head of the trail, just off Interstate 26 at the Unicoi Road exit. Information signs have also been installed. As of early 2014, plans were still being made to build an arts and cultural heritage center, known as Tanasi, near the trailhead.

Only three federally managed fire towers remain in the northern Cherokee National Forest, including Meadow Creek in Cocke County, Camp Creek Bald in Greene County and Pinnacle Mountain. Several organizations and historic property registers have been established to help protect and preserve the iconic structures. The Forest Fire Lookout Association was founded in 1990. Its mission is to research current and former forest fire lookout sites, ground cabins and early forest fire detection methods. It promotes the protection, enjoyment and understanding of lookouts.

Members of the FFLA include lookout enthusiasts, hikers, conservationists, forest fire personnel, foresters, writers, journalists and members of the environmental community. The FFLA, according to its website, has several duties. It organizes lookout restorations, provides grants for lookout projects, documents lookout sites, provides historical research of lookouts, assembles programs to increase lookout awareness and documents anecdotes of lookout operators, both past and present. The FFLA holds conferences twice annually in the eastern and western United States and Canada. There is also a board of directors, as well as state, regional and local chapters throughout North America. Each chapter has a director, including Peter Barr, who has organized the Save Shuckstack campaign in Swain County.

The FFLA first formed in early 1990 in the District 17 garage at French Creek State Park in Pennsylvania. Afterward, the group toured the nearby Hopewell Fire Tower and lookout cabin. The organization eventually spawned the official FFLA. By 2014, the FFLA had become the United States' largest and most active organization geared toward forest tower preservation.

The National Register of Historic Places and National Historic Lookout Register recognize fire lookout towers across the United States. Listing in the National Historic Lookout Register generally precedes listing in the National Register of Historic Places.

Annie McDonald, a preservation specialist for the western region of North Carolina, said that two properties in her state have been considered for inclusion in the NRHP. The towers, Mount Sterling and Shuckstack, are both located on the North Carolina side of the Great Smoky Mountains National Park. The sites have been placed on North Carolina's National Register Study List, and both were listed at the national park's request.

"Study List designation is the first step to listing in the National Register of Historic Places for properties in North Carolina," McDonald explained. "While this designation does not automatically mean that a property is eligible for listing in the National Register of Historic Places, it is a good indicator that the property may be historically significant enough and retain a sufficient degree of physical integrity to warrant consideration for this honor."

Each state has different processes for listing in the NRHP. In North Carolina, the program is operated by the state's Historic Preservation Office. In Tennessee, the state's Historical Commission maintains the registry. In Virginia, it's the Department of Historic Resources. The State Historic Preservation Office maintains South Carolina's nominations. The Historic Preservation Division operates Georgia's nominations. It is recommended to contact each state's historical preservation office to determine how to nominate a property; however, every state's processes are in compliance with the national register.

For a building or structure to be eligible for listing in the NRHP, it must be considered to be historically significant either for its representation of broad patterns of history or events; because of its association with an individual or group that has had a lasting impact on the history of the community or state; or because it represents a type, period or method of construction. If a site yields information important in history or prehistory, it may also be placed on the NRHP.

Once a site, such as a fire tower, is determined to be of historic significance, it must be evaluated to ascertain whether or not it retains sufficient physical and architectural integrity to the historic period. It must pass both aspects before inclusion on the NRHP. "Yes, fire towers may qualify for National Register designation, but they must go through this evaluation process through consultation with our office," McDonald said.

A search of the NRHP database in March 2014 found twenty-one fire lookouts. All were in the western and midwestern states. The beautiful stone fire tower on Chenocetah Mountain in north Georgia was the only lookout listed on the register in the Blue Ridge Mountains. It was added in 1984. A handful of other towers have been considered for recognition, including Mount Sterling, Shuckstack and Mount Cammerer in the Great Smoky Mountains National Park.

In 2005, the Pisgah and Nantahala National Forests contracted Edwards-Pitman Environmental to study and evaluate the twelve remaining lookouts in western North Carolina's national forests. Part of the survey was to determine whether the lookouts were eligible for the NRHP. According to the report, three major hurricane and rain events affected the lookout towers in the fall of 2004. As a result, the U.S. Forest Service also wanted to determine the effects of repairs and stabilization of the lookouts.

During the study, Edwards-Pitman architectural historians visited each tower site to document the tower and its setting with photographs, maps, sketches and field notes. According to the survey, all but two of the twelve remaining lookouts appear to be eligible for the NRHP. Prior to the survey, the state had already recommended listing of Panther Top in Cherokee County and Wauchecha Bald in Graham County. Both the state and the survey have decided that the lookouts at Camp Creek Bald and Wesser Bald are not eligible for the survey because of major changes to the structures. The forest service replaced Camp Creek Bald's original square cab with a circular cab, and Wesser Bald's original cab was destroyed by fire and replaced with an observation deck.

The survey noted that the twelve fires lookouts appear to eligible for the register because of their association with forest conservation efforts in the early and middle twentieth century. And despite changes to their designs and replacement materials from ongoing maintenance, the survey concluded that the lookouts retain enough of their overall character and design to be considered for the NRHP with the exception of Camp Creek Bald and Wesser Bald.

In addition, the survey noted that two of the stone towers, Wayah Bald and Yellow Mountain, appear to be eligible for their design. Although Wayah Bald was altered in 1947, the survey said that those changes occurred more than fifty years ago and contribute to the historic significance of the tower.

One tower in eastern North Carolina, the Warren County Fire Tower, is already on the NRHP. The Warren County lookout is a one-hundred-foot Aermotor tower and was constructed in 1932.

In 1986, eight fire lookouts in Glacier National Park were included in the NRHP. The park built the towers as a chain of lookouts to monitor forest fires starting in the 1920s. All eight towers—Apgar Fire Lookout, Heaven's Peak Fire Lookout, Huckleberry Fire Lookout, Loneman Fire Lookout, Mount Brown Fire Lookout, Numa Ridge Fire Lookout, Scalplock Fire Lookout and Swiftcurrent Fire Lookout—were constructed between 1923 and 1945.

Nearly one thousand sites around the country are listed on the National Historic Lookout Register, a cooperative effort of the U.S. Forest Service, the Forest Fire Lookout Association, state agencies, private organizations and private individuals. The listing recognizes individual lookouts that have helped in fire detection and wildfire management for at least fifty years.

According to the NHRL, the listing is often the first step toward nomination to the NRHP, which is maintained by the U.S. Department of the Interior. Unfortunately, structural modifications to the fire towers result in the rejection of the structures to the NRHP. In many cases, towers only qualify for listing in the NHRL due to modifications.

Through these organizations and registers, restoration and preservation projects are under consideration or ongoing throughout the southern Appalachian Mountain region. Despite a lack of government funds to restore fire lookout towers, several of the iconic structures are being saved by concerned citizens and organizations. Without these projects, several towers continue to be razed, dismantled and removed from the mountains.

BIBLIOGRAPHY

Adams, Daniel. *Methods and Apparatus for the Prevention and Control of Forest Fires, as Exemplified on the Arkansas National Forest*. U.S. Department of Agriculture. Washington, D.C.: National Archives and Records Administration, 1912.

Barden, Linda. "Rabun's Forest Rangers." Available at Rabun County Historical Society. Clayton, Georgia, 2009.

Barr, Peter. *Hiking North Carolina's Lookout Towers*. Winston-Salem, NC: John F. Blair, Publisher, 2008.

———. "Save Shuckstack." Peter on the Appalachian Trail. http://peterontheat.com/save-shuckstack.

Bauer, Jennifer. *Roan Mountain: A Passage of Time*. Winston-Salem, NC: John F. Blair, Publisher, 1991.

Bristol Herald Courier. Various articles on High Knob Fire Tower, 2007–2014.

Brooks, Bob. "Outdoors for Tomorrow." *Boys Life* (December 1951).

Carter, Kayla. *Civilian Conservation Corps in Unicoi County Pioneered East Tennessee Reforestation and Preservation*. Johnson City, TN: Johnson City Press, 2013.

Davis, Frank C. *My C.C.C. Days: Memories of the Civilian Conservation Corps*. Boone, NC: Parkway Publishers Inc., 2006.

Devlin, Sherry. "Taming the Dragon." *Missoulian*, August 20, 2000.

Dyer, Delce. *Rural Historic Landscapes and Interpretive Planning on Southern National Forests*. National Park Service, 2014.

Eastern Lookouts. http://easternuslookouts.weebly.com.

Edwards-Pitman Environmental Inc. *Lookout Towers Documentation and National Register of Historic Places Evaluation*. Durham, NC: self-published, 2005.

Erwin Record. Various articles on Pinnacle Mountain Fire Tower, 2008–2011.

Forest Fire Lookout Association. http://www.firelookout.org.

The Forest History Society. "The 1910 Fires." http://www.foresthistory.org/ASPNET/Policy/Fire/FamousFires/1910Fires.aspx.

———. "U.S. Forest Service History." http://www.foresthistory.org/ASPNET/USFHSHome.aspx.

———. "U.S. Forest Service History: The Weeks Act." http://www.foresthistory.org/ASPNET/Policy/WeeksAct/index.aspx.

French, Madrid. "Mission 66." http://mission66.com.

Ingle, Laura. "Every Day Is Fire Day: A Study of Historic Fire Towers and Lookout Life in the Great Smoky Mountains National Park." MA thesis, Clemson University, 2011.

Johnson City Press. Various articles on Pinnacle Mountain Fire Tower, 2008–2014.

Jolley, Harley E. *The CCC in the Smokies*. Gatlinburg, TN: Great Smoky Mountains Association, 2001.

Jurgelski, William. "Burning Seasons, Burning Bans: Fire in the Southern Appalachian Mountains, 1750–2000." *Appalachian Journal* (2008).

Kestner, Jack. *Fire Tower*. Emory, VA: Clinch Mountain Press, 2007.

———. *A View from the Mountain*. Emory, VA: Clinch Mountain Press, 2006.

Kingsport Times News. Various articles on High Knob Fire Tower, 2007–2014.

Knoxville News-Sentinel. Various articles, 2008–2014.

Kresek, Ray. "Osborne Fire Finder." U.S. Forest Service, 2014. http://www.fs.usda.gov/detail/r1/fire-aviation/?cid=stelprdb5108647.

Lohmann, Bill. "Lohmann: A Mountain, a Family and a Tower that's a Little Too Tall for Some of Us." *Richmond Times Dispatch*, June 18, 2014.

Mastran, Shelley Smith. *Mountaineers and Rangers: A History of Federal Forest Management in the Southern Appalachians.* McLean, VA: U.S. Department of Agriculture, 1983.

Mountain Retreat & Learning Center. "History of Little Scaly Mountain." Highlands, North Carolina.

National Historic Lookout Register. http://www.nhlr.org.

National Park Service. *Cone Park Heritage Trails*. Blue Ridge Parkway, 2014.

Pigeon, Resin E., assistant regional forester. *Holston Mountain Fire Report*. Forest Service, Atlanta, Georgia, October 16, 1943.

Pinchot, Gifford. *The Fight for Conservation*. Seattle: University of Washington Press, 1910.

Rabun County Historical Society. "High Elevation History." *Vintage Rabun Quarterly* (December 2012).

Silcox, Ferdinand. *Fire Prevention and Control of the National Forests*. U.S. Department of Agriculture. Washington, D.C.: National Archives and Records Administration, 2014.

South Carolina Forestry Commission "South Carolina Forestry Commission History and Archives." http://www.state.sc.us/forest/facts.htm.

Spencer, Marci. *Pisgah National Forest: A History*. Charleston, SC: The History Press, 2014.

U.S. Department of the Interior. "Chenocetah." National Register of Historic Places nomination form.

———. "Clingmans Dome." National Register of Historic Places nomination form.

U.S. Department of the Interior, National Park Service. *Lookout Manual*. Great Smoky National Mountain Park Library, 1940.

———. "Wildland Fire History." http://www.nps.gov/applications/fire/wildland-fire.

INDEX

D

E

F

G

H

ABOUT THE AUTHOR

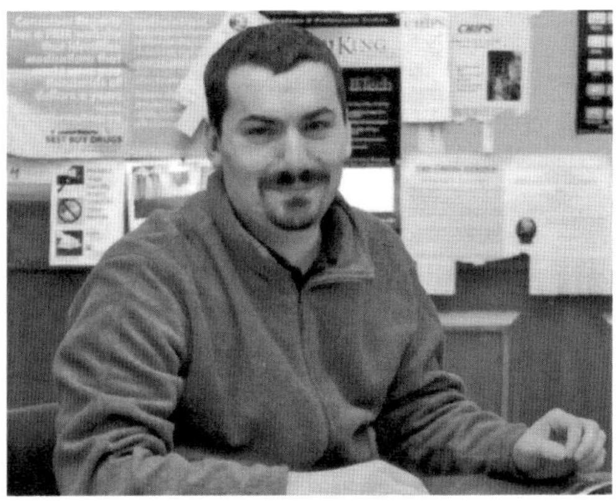

A newspaper journalist and author for more than a decade, Robert Sorrell is a graduate of East Tennessee State University. He has worked part time and full time as a newspaper reporter at the *Erwin Record*, the *Elizabethton Star* and the Pulitzer Prize–winning *Bristol Herald Courier*. In addition, Robert has authored numerous stories for regional and national magazines, including *Grit* magazine, *Georgia* magazine and *Blue Ridge Country*. He is fascinated with the history and heritage of the southern Appalachian Mountains. During his studies of the mountains, he has fallen in love with the region's fire lookout towers.